GENERAL, ORGANIC, AND BIOLOGICAL CHEMISTRY

A GUIDED INQUIRY

MICHAEL P. GAROUTTE
Missouri Southern State University

BICENTENNIAL 1807 WILEY 2007 BICENTENNIAL

JOHN WILEY & SONS, INC

BS

To order books or for customer service please, call 1-800-CALL WILEY (225-5945).

ISBN 0-471-76359-4

Printed in the United States of America

10 9 8 7 6 5 4 3 2 1

12/28/06

To Susan,
who put up with me while I was writing this book,
and to my daughters
Audrey and Madeleine.

To the Instructor

The activities presented in *General, Organic, and Biological Chemistry: A Guided Inquiry* have been written or adapted to support process-oriented guided-inquiry learning (POGIL) in an allied health or GOB (General-Organic-Biological) chemistry course. Current educational research shows that most students experience improved learning when they are actively engaged, and when they are given the opportunity to construct their own knowledge. POGIL activities are designed for use with self-managed teams that employ the instructor as a facilitator of learning rather than as a source of information.

Each activity in this set may be used independently, but some questions may depend upon topics examined in earlier activities. These activities do not replace a traditional textbook, but rather enhance its use. Any standard text may be used, and you are encouraged to correlate reading and/or homework assignments from the text with the ChemActivities in this book. Many instructors will choose to teach using POGIL during every class meeting. If you are new to guided inquiry, you might select one activity per week to introduce more student participation into your class.

ChemActivities 3, 7, 10, 13, 15, 16, 19, and 21 are adapted from activities published in Moog, R.S. and Farrell, J.J. *Chemistry: A Guided Inquiry,* 2nd and 3rd editions, Wiley, 2002 and 2006.

An *Instructor's Guide*, available as a PDF download from Wiley.com, contains instructor notes for each activity, example quizzes, suggested answers for the Critical Thinking Questions and solutions to the Exercises.

Anyone who teaches a GOB nonmajors course will know that, from the point of view of a chemist, the course is all about compromises. This set of activities is no exception. I tell my students in this course that it is five semesters of chemistry compressed into one. Some of you may have a two-semester GOB sequence. Regardless, choices must be made about which topics are and are not important, and about the depth to which selected topics are investigated. I have attempted to focus here on a core set of skills that will help students succeed in their future courses and careers. This has meant, for example, much less emphasis on traditional organic chemistry reactions than is present in many current textbooks.

For more information about the POGIL project and guided-inquiry learning, including a list of available teacher resources, materials to use in the classroom, and references to relevant research, please visit the POGIL Web site at http://www.pogil.org.

Feedback regarding the effectiveness of the materials and suggestions for improvements would be appreciated. Send this assessment information to the author by email (address available from the Missouri Southern State University Web site).

To the Student

Many of you may see this chemistry class as simply a requirement to be met so that you can get on with the education you "really need" to begin your career. And many of you have been very successful in your academic careers by listening to lectures and performing well on exams. When you found out (or perhaps are finding out right now) that in this course, you will be spending much class time doing "group work," you may have felt that someone placed a hurdle in your path. This class is different.

This class is not simply about "how to do the calculation," nor is a major goal of this class satisfied by simply learning what numbers to plug into what equation—though those things are good to know. The main goal is to understand what the numbers *mean*.

In a class such as this, you may be frustrated at times because you cannot immediately see the "right" answer to a question. Actually, it is by design that some answers are not immediately obvious. Sometimes you will write an answer and go on to a later question, only to find that the later question causes you to reevaluate the earlier question. This is OK! Later, when you have the "aha!" moment, you will not easily forget what you have learned.

The most common reason that students feel their group is "not working" is that one member is contributing either too much or too little. It is each group member's responsibility to make sure that all group members contribute. Do not hesitate to talk to your instructor if you feel that your group needs help learning how to be more productive.

Here is the best single suggestion, given by many instructors and students, for how to succeed in and enjoy this course: find a study partner or group, and meet regularly! When you have study partners, you have a reason to be prepared (they are counting on you), and if you can't come up with the answer together, then you are less shy about asking the instructor—if none of you can figure it out, it must need some teacher clarification!

Hopefully in this class, you will learn that chemistry is actually relevant to your career choice, and that studying it can be fun as well! If you have any suggestions for improving this book, please email me at my university.

Michael Garoutte
Missouri Southern State University
July 2006

Some comments from former students in this course

I thought that I wouldn't like this class but it has been really interesting to come to class every day and actually learn something that pertains to life.

If you work well with others and are able to learn as a group and be challenged by your group of peers to strive to really learn the subjects then I recommend this class. But if you'd rather work on your own and not get help from others then this class would be of no benefit to you.

This is one of the hardest classes I have ever taken. But was the only one that taught me to seek out the answer instead of having it handed to you. This class will help me in future classes, because I have gained good study skills. For that I wish to thank [the instructor] :)

I think an open mind and patience for the first few weeks can best describe my recommendation to other student's wanting to take this class... I didn't like this class at first, but it grew on me and I learned more than I thought I would have. So, I will recommend this class to other students.

When I took chemistry in High School, I did not understand or remember a thing. Now that I have taken this class, I LOVE CHEMISTRY! I really like learning things on my own and not just someone talking at me. I feel that I have made a TREMENDOUS improvement in the field of Chemistry, and just makes me further appreciate the subject.

Acknowledgements

Thanks to Richard L. Schowen, Professor Emeritus, University of Kansas, who was my graduate advisor, and who after reading the first draft of my dissertation, praised my writing style and stated that I might want to consider authoring a textbook someday. Dick, your statement gave me the confidence to undertake this project. This isn't a textbook, but I hope it meets with your approval.

Thanks to Andrei Straumanis, College of Charleston, and Renée Cole, Central Missouri State University, who introduced me to POGIL at a workshop in 2003, and who have provided much advice and encouragement.

Thanks to Rick Moog and Jim Spencer of Franklin & Marshall College, who didn't laugh when I showed them the first draft of these activities, and who apparently believed in the adage that "if you can't say something nice, don't say anything at all." Special thanks to Rick, who has continued not to laugh as I have repeatedly barraged him with questions about, well, everything, and who reviewed some of these activities and gave suggestions for their improvement. Thanks also to John Farrell, who along with Rick, wrote the book that inspired this one.

Thanks to all the people of the POGIL project, who have kindled my enthusiasm by sharing their ideas and materials, by listening, by allowing me to contribute, and by continuing to invite me to share my ideas. I can't imagine a group of more helpful and supportive individuals.

Thanks to my colleagues and department head at Missouri Southern, who have supported my choice to "try something different" in class. Many teachers don't have it so good. Thanks especially to Mel Mosher, who gave me advice when I was teaching this course for the first time; and to Marsi Archer who graciously agreed to use an earlier version of these materials in a course in 2004, and who provided valuable feedback.

Thanks to Deborah Edson and the folks at Wiley who agreed to publish this book, and who have been patient with all my questions.

Thanks to all my students at Missouri Southern State University, who have helped me "beta-test" various versions of these materials.

And finally, many thanks to my family—parents, in-laws, wife, and children—who have all contributed in some way by making it possible for me to spend time working on this book. I want all of you to know that I haven't forgotten you, and that your sacrifices are appreciated.

Table of Contents

Recorder's Report A

Date ____________________ ChemActivity ______________

Manager ____________________________ Recorder ____________________________

Presenter ____________________________ Reflector ____________________________

Group Assessment: /10

One thing we learned today is:

One concept we need to work on is:

Other Comments:

Recorder's Report B

Date ____________________ ChemActivity ______________

Manager ____________________________ Recorder ____________________________

Presenter ____________________________ Reflector ____________________________

Group Assessment: /10

Comments:

Important ideas, concepts, results, *etc.*

Working in Groups; Estimation

Information: Brief description of roles

Much of the class time in this course will be spent working in groups of three or four. Each member of the group will be assigned to a particular role. Some typical roles (and their descriptions) are listed below. Your instructor will let you know how the roles will function in your course.

Manager: Manages the group. Ensures that members are fulfilling their roles, that the assigned tasks are being accomplished on time, and that **all members of the group participate together** in activities and **understand the concepts**.

Presenter: Presents the work of the group to the entire class when called upon. The presenter may be called to the board to write out and explain the group's solution to a problem. Frequently the instructor will ask what the group responded to a particular question or whether the group agrees with another group's response. It is the presenter's role to reply to these questions.

Recorder: Records (on report form) the names of each of the group members at the beginning of each day. Keeps track of the group answers and explanations, along with any other important observations, insights, *etc.* The completed report with answers to any questions asked may be submitted to the instructor at the end of the class meeting.

Reflector or Strategy Analyst: Observes and comments on group dynamics and behavior with respect to the learning process. For example, the Reflector might comment that a particular group member is dominating the discussion.

If a group member is absent, then one member may have to fulfill more than one role.

Model 1: A centimeter ruler

0 1 2 3 4 5 6 7 8 9 10 11 12 13 14 15

Critical Thinking Question:

1. Estimate the number of ping-pong (table tennis) balls that would completely fill this classroom. First, decide upon a "plan of attack" as a group. You may or may not choose to use the centimeter ruler. For the purposes of this exercise, you may assume that the room is rectangular in shape and that it is completely empty of desks, people, *etc.* You may get up and move around the room. When your group has an answer, the recorder may be asked to write it on the board.

Exercise:

1. Read the assigned pages in the text, and work the assigned problems.

Types of Matter; Chemical and Physical Changes

(How can we classify matter?)

Model 1: Examples of various types of matter at room temperature

Item	Classification	State (or states)	Formula (or formulas)
aluminum	element	solid	Al(s)
hydrogen	element	gas	H_2(g)
water	compound	liquid	H_2O(l)
table salt	compound	solid	NaCl(s)
salt water	homogeneous mixture	aqueous solution	H_2O(l) and NaCl(aq)
coffee ("black")	homogeneous mixture	aqueous solution	H_2O(l) and many others
muddy water	heterogeneous mixture	liquid + solid	H_2O(l) and other stuff

Critical Thinking Questions:

1. Consider Model 1. How does the formula of an *element* differ from that of a *compound*?

2. How can you distinguish elements from compounds based on their chemical formulas?

3. How does a pure substance (*i. e.,* element or compound) differ from a mixture? Describe.

4. Hypothesize on the meaning of the labels (s), (l), (g), and (aq) on the formulas.

Information: States of matter

Matter can be classified by its physical **state** (or **phase**): **solid, liquid,** or **gas**. Most substances can exist in each of the three states depending on temperature and pressure. For example, H_2O is normally a liquid at room temperature, ice at temperatures below 0°C, and vapor at temperatures above 100°C. Although we think of iron as a solid, it can be melted and even vaporized if the temperature is high enough. Changes between these states are usually considered physical—not chemical—changes, since the chemical formula is the same.

The **phase labels** (s), (l), or (g) can be written after a formula to signify the physical state. So, $H_2O(g)$ would mean gaseous water, *i. e.,* water vapor.

Critical Thinking Questions:

5. Describe what is happening in this process: $H_2O(l) \rightarrow H_2O(s)$

6. Would the process in **C**ritical **T**hinking **Q**uestion **5** (**CTQ 5**) be considered a chemical or a physical process? Explain.

Information: Classifications of matter

Chemistry is the science that deals with matter and the changes matter undergoes. Matter can be divided into two main types: pure **substances** and **mixtures** of substances.

A **substance** cannot be separated into other kinds of matter by physical processes such as filtering or evaporation, and is either an **element** (*e. g.,* aluminum) or a **compound** (*e. g.,* H_2O). Compounds are made of two or more elements chemically combined. The elements themselves cannot be separated into simpler substances even by a chemical reaction.

On the other hand, **mixtures** can be separated by physical means. Mixtures that have the same composition throughout are called **homogeneous** (*e. g.,* salt water); those that do not are called **heterogeneous** (*e. g.,* Italian salad dressing).

Atoms of a single element can exist as individual atoms (*e. g.,* aluminum, Al) or in chemical combination to form molecules (*e. g.,* hydrogen, H_2). When atoms of two or more elements combine chemically they form **compounds** (*e. g.,* water, H_2O). Subscripts following an element in the formula represent the number of atoms of that element in the formula.

Model 2: Flow chart for classifying matter

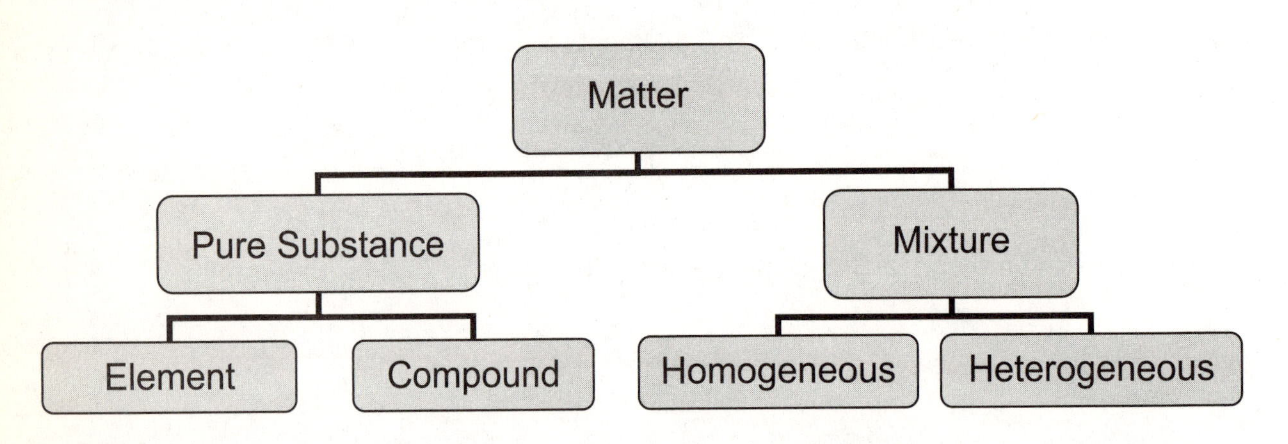

Model 3: Some different representations of the water molecule

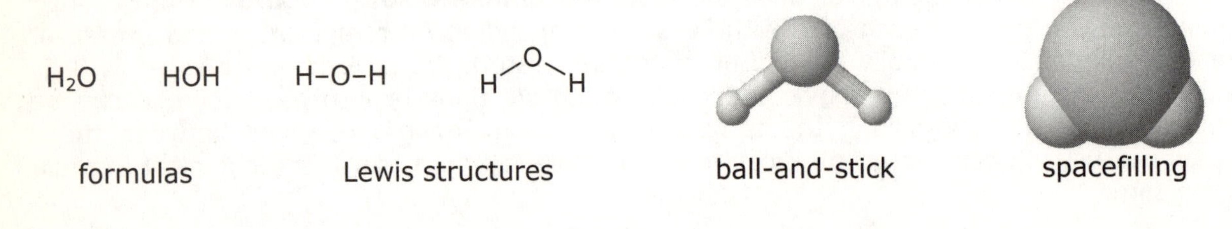

Critical Thinking Questions:

7. Look at a periodic table of the elements. About how many elements are known?

8. Approximately how many elements are metals? (Estimate, don't count!)

9. There are **thousands** of organic compounds known—compounds formed out of only a few different elements (carbon, hydrogen, oxygen, nitrogen). Explain how this can be possible.

Exercises:

1. Write the formula of each molecule for which the ball-and-stick structures are shown.

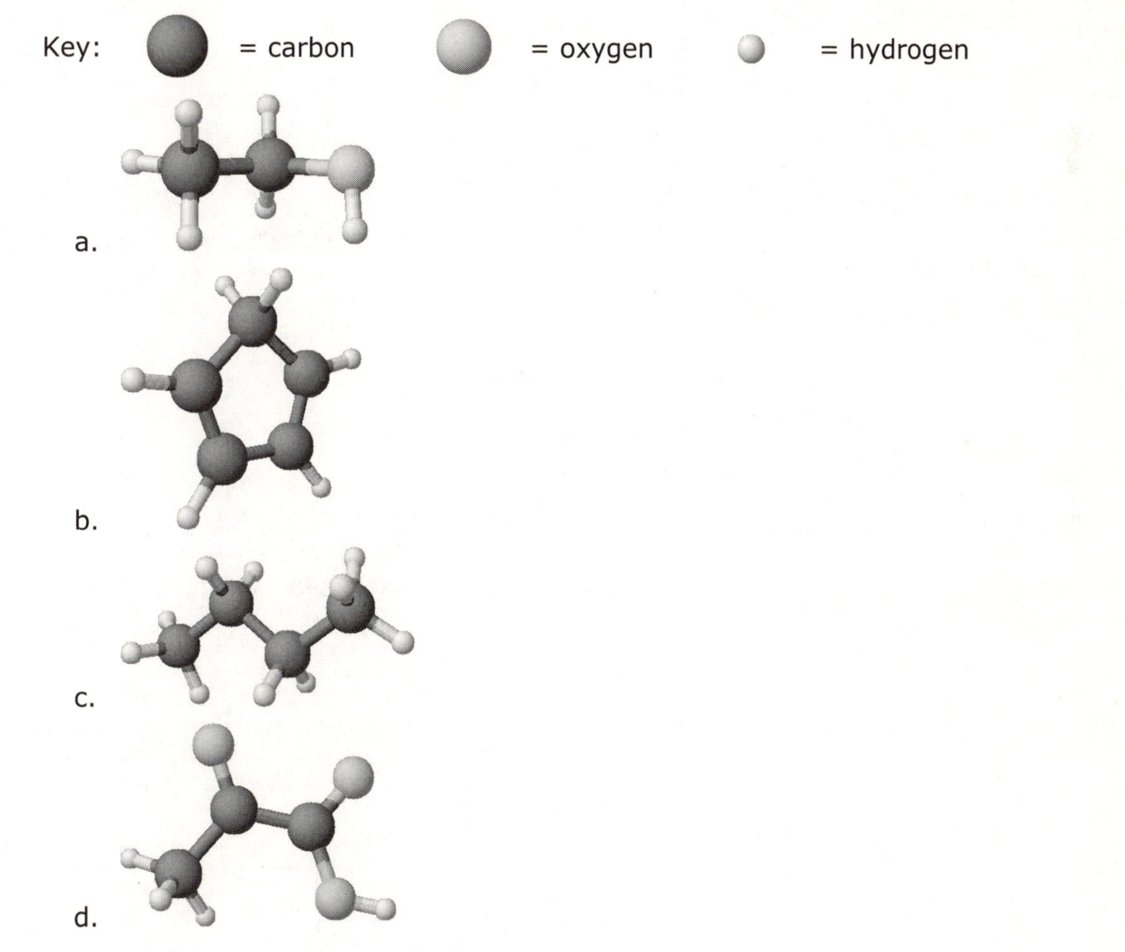

2. Using a periodic table, identify the elements represented in each formula, and state the number of atoms of each element in the formula. The first one has been done for you.

    a. NH_3 (ammonia) *one nitrogen atom, three hydrogen atoms*

    b. $C_6H_{12}O_6$ (glucose)

    c. $Mg(OH)_2$ (milk of magnesia)

    d. H_2SO_4 (sulfuric acid, "battery acid")

    e. $C_{17}H_{18}F_3NO$ (fluoxetine, Prozac)

3. Using the flow chart in Model 2 to help you, classify each of the following as either a mixture or pure substance. For each **substance**, tell whether is it an **element** or a **compound**. For each **mixture**, tell whether it is **homogeneous** or **heterogeneous**; then list two or more components of the mixture.

    a. a lead weight

    b. apple juice

    c. baking soda ($NaHCO_3$)

    d. air

    e. a 14-karat gold ring

    f. a 24-karat gold coin

    g. helium in a balloon

    h. beach sand

    i. concrete

    j. whole blood

    k. carbon dioxide

4. In the space below, draw a picture of three water molecules in the ball-and-stick representation.

5. Which of the choices below (I or II) would best represent the three molecules you drew in Exercise 4? Explain your choice.

Choice I	Choice II
H_6O_3	$3\ H_2O$

6. Learn the **names** and **symbols** of the elements your instructor suggests. A good starting point is the first 30 elements, plus Br, Sr, Ag, Sn, I, Ba, Pt, Au, Hg, Pb. Spelling counts! You do **not** need to memorize any **numbers**, as a periodic table will always be available for your use.

7. Read the assigned pages in the text, and work the assigned problems.

Atoms and The Periodic Table*

(What are atoms?)

Model 1: Schematic diagrams for various atoms

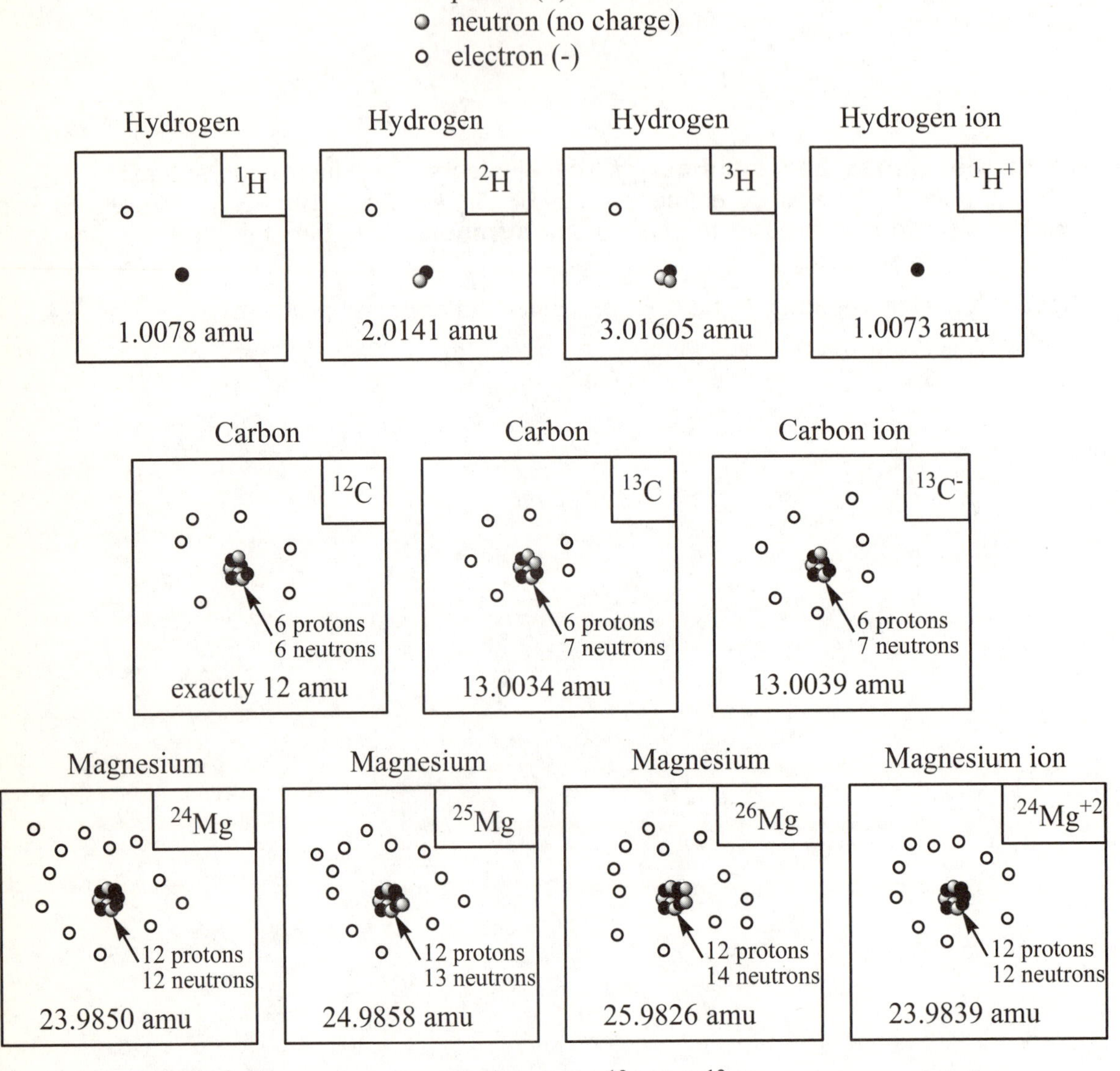

^{1}H, ^{2}H and ^{3}H are **isotopes** of hydrogen. ^{12}C and ^{13}C are **isotopes** of carbon.

^{12}C and ^{13}C may also be written as "carbon-12" and "carbon-13"

The **nucleus** of an atom contains the protons and the neutrons (if any).

* Adapted from ChemActivity 1, Moog, R.S.; Farrell, J.J. *Chemistry: A Guided Inquiry*, 3rd ed., Wiley, 2006, pp. 2-5.

Critical Thinking Questions:

1. Look at the Schematic Diagrams for carbon. What do all three carbon atoms (and ions) have in common?

2. What do all four hydrogen atoms and ions have in common?

3. What do all magnesium atoms and ions have in common?

4. Look at a periodic table (for example, in your text). Considering your answers to CTQs 1-3, what is the significance of the **atomic number**, above each element in the table?

5. How many protons are in all chlorine atoms?_____ Do you think chlorine atoms exist with 18 protons? _____ Why or why not?

6. What is the significance of the superscripted + or – symbols to the right of the symbols
   a) for the hydrogen ion?

   b) for the carbon ion?

   c) Write one complete sentence to define the term ion.

   d) A positively charged ion is called a cation (pronounced "cat ion"), and a negatively charged ion is called an anion. Which term applies to the magnesium ion?

7. In a box in the corner of each schematic in Model 1 is the element symbol and the mass number for the atom (superscript on the left side of the element symbol).
   a) How is this mass number determined?

   b) Why is it called a "mass" number?

8. What is the mass number for the following atoms:

   a) ^{37}Cl _____ b) ^{238}U _____ c) carbon-12 _____ d) carbon-13 _____

9. What structural element do all isotopes of an element have in common? How are their structures different?

10. Considering what you know about isotopes, do all atoms of an element weigh the same? Explain.

Model 2: The periodic table

Look at a periodic table (*e. g.*, in your textbook).

For each element, the **atomic number** is shown above the element symbol, and the average **atomic mass** (or **atomic weight**) is shown underneath. Since the table will always be available, there is no reason to memorize either of these numbers.

There are 18 columns, known as **families** or **groups**. The groups can be numbered in various ways. We will use the numbers 1-18 for group designations. Elements in the same group often have similar properties. Some groups have family names.

Your table may be divided into the metals, semimetals, and nonmetals. If not, note that a "stair-step" line which starts between boron (B) and aluminum (Al) roughly divides metals from nonmetals. Semimetals (or "metalloids") lie along the line.

There are seven periods, or rows, in the table. The two rows of 14 elements at the bottom actually fit between groups 2 and 3 on the main table. They are shown at the bottom because that way, they "fit" on the page better.

Critical Thinking Questions:

11. Are most elements metals or nonmetals? ______________________

12. From your experience, list some properties of metals.

13. From your experience, list some properties of nonmetals.

14. Transistors and computer "chips" are made of semiconductor materials. What kind of elements would be used for this purpose? Give an example.

Exercises:

1. Fill in the following table.

Symbol	Atomic Number	Mass Number	Number of Protons	Number of Neutrons	Number of Electrons
^{40}K					
			15	17	18
Zn^{2+}				38	
	35	81			36

2. Complete the table below, classifying each element as a metal, nonmetal, or semimetal. Watch spelling!

Symbol	Name	Classification
Pd		
Cl		
	germanium	

3. Complete the table below, classifying each element as an alkali metal, alkaline earth metal, transition metal, halogen, or noble gas. Watch spelling!

Symbol	Name	Classification
F		
Ca		
	rubidium	
Cr		
	krypton	

4. Learn the family names of groups 1, 2, 17, 18.

5. Learn to recognize which elements are metals, nonmetals, semimetals (metalloids), transition elements or main group elements.

6. Read the assigned pages in the text, and work the assigned problems.

Unit Conversions: Metric System

(*What is a conversion factor?*)

Model 1: Fuel efficiency of a particular automobile

A particular automobile can travel 27 miles per gallon of gasoline used.

The automobile has a 12-gallon gasoline tank.

At a particular location, gasoline costs $3.00 (3.00 USD) per gallon.

Critical Thinking Questions:

1. Three statistics are given in Model 1. Circle the two statements that give numerical **ratios**.
2. One statement in Model 1 gives a measured **quantity**. Write the quantity (with the associated **unit**).
3. Write each ratio that you circled in Model 1 as a **fraction**. Your fraction should have a **number** and a **unit** in both the numerator and the denominator of the fraction.
4. How many miles can the automobile travel on a full tank of gasoline? Show your work by writing the **quantity** from CTQ 2 multiplied by the appropriate **fraction** from CTQ 3. Show all units.
5. Explain why the **answer** to CTQ 4 does not include the unit "gallons."
6. Explain why the **fraction** used in CTQ 4 may be called a **conversion factor**.
7. Do all four conversion factors below give equivalent information? Explain your answer.

$$\frac{27\ \text{mi}}{1\ \text{gal}} \qquad \frac{27\ \text{mi}}{\text{gal}} \qquad \frac{1\ \text{gal}}{27\ \text{mi}} \qquad \frac{\frac{1}{27}\ \text{gal}}{\text{mi}}$$

Model 2: Definitions of the inch and the foot

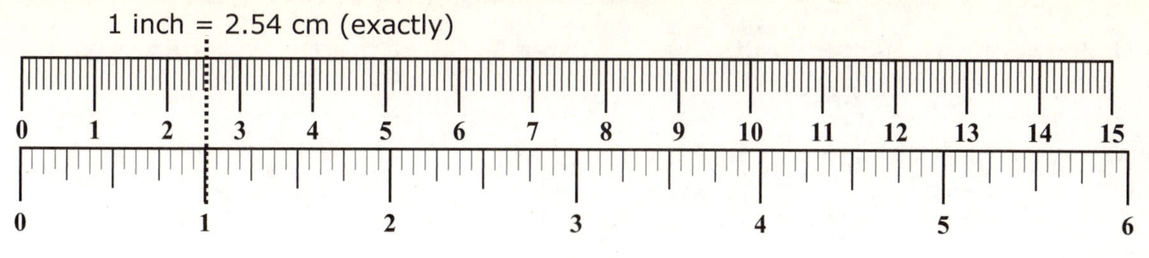

There are exactly 12 inches in one foot.

Critical Thinking Questions:

8. How many centimeters are in one inch?

9. Draw a large X through each **incorrect** conversion factor below.

$$\frac{1\text{ cm}}{2.54\text{ in}} \qquad \frac{2.54\text{ cm}}{1\text{ in}} \qquad \frac{2.54\text{ in}}{1\text{ cm}} \qquad \frac{1\text{ in}}{2.54\text{ cm}}$$

10. Suppose you want to convert a height from inches into centimeters. Circle the conversion factor in CTQ 9 that you would use. Explain your choice.

Model 3: The unit plan

A unit plan begins with the **unit of the known quantity** and shows how the units will change after multiplying by each conversion factor used, in order. The unit plan for CTQ 10 would be:

$$\text{in} \rightarrow \text{cm}$$

Each arrow in the unit plan represents one conversion factor.

Critical Thinking Questions:

11. A basketball player is seven feet tall.

    a. Using Models 2 and 3 for reference, complete the unit plan for converting the height of the basketball player into centimeters.

    feet →   →

    b. Write the two conversion factors corresponding to each arrow in part (a).

    c. Perform the calculation by multiplying the **quantity** by each **conversion factor** in order. Show all work with units.

Information: Metric system units and prefixes

Table 1: Some metric system units

Quantity	Unit	Symbol
distance	meter	m
time	second	s
mass	gram	g
volume	liter	L
temperature	kelvin	K

Table 2: Some metric system prefixes

Prefix	Symbol	Factor	Meaning
mega	M	10^6	one million
kilo	k	10^3	one thousand
deci	d	10^{-1}	one tenth
centi	c	10^{-2}	one hundredth
milli	m	10^{-3}	one thousandth
micro	μ	10^{-6}	one millionth

The International (metric) System prefixes can be associated with any unit. So for example, one can write millimeters (mm), milliseconds (ms), milligrams (mg), or milliliters (mL).

There are more prefixes than these. For example, personal computer clock speeds are measured in gigahertz (1 GHz = 10^9 cycles per second), and the MCV (Mean Cell Volume) of red blood cells is measured in femtoliters (1 fL = 10^{-15} liters). But we will stick to the most common ones. **You will need to memorize the information in Tables 1 and 2.**

Critical Thinking Question:

12. Circle the unit that would be commonly used to measure the indicated item.

a.	The volume of liquid in a can of Coca-Cola:	μL	mL	L	kL	ML
b.	The mass of a human being:	μg	mg	g	kg	Mg
c.	The mass of aspirin in one tablet:	μg	mg	g	kg	Mg
d.	The volume of the gasoline tank in a Volkswagen	μL	mL	L	kL	ML

Model 4: Conversion factors from metric system prefixes

The prefixes from Table 2 give you an **equality** *to the base unit (the unit with no prefix).* For example:

$1\text{ mm} = 10^{-3}\text{ m}$ $1\text{ ks} = 10^3\text{ s}$ $1\text{ cg} = 10^{-2}\text{ g}$ $1\ \mu\text{L} = 10^{-6}\text{ L}$

The equality can be used to create a **conversion factor**. For millimeters, the possible conversion factors are*:

$$\frac{1\text{ mm}}{10^{-3}\text{m}} \text{ and } \frac{10^{-3}\text{ m}}{1\text{ mm}}$$

Example Problem:

How many mm are in 1.89 m?

The unit plan is: m → mm

Using the **first** conversion factor in Model 4, $\frac{1\text{ mm}}{10^{-3}\text{m}}$, will cause the meters (m) to cancel out, so we should choose that one. Then do the conversion:

$$1.89\text{ m} \times \frac{1\text{ mm}}{10^{-3}\text{m}} = 1890\text{ mm}$$

Remember, to enter 10^{-3} into most simple scientific (non-graphing) calculators, use the **exponent** (EE or EXP) key (*e. g.*, press 1 EE 3 ±). Do not use the **power** (y^x) key.

* If you have already memorized equalities such as "1 km = 1000 m" and "1000 mm = 1 m," you may use them. However, this book will always use the powers of 10 and prefixes shown in Table 2.

Critical Thinking Question:

13. How many centimeters are in 2.2045×10^{-2} kilometers?

   a. Complete the unit plan.

   km $\rightarrow$ $\rightarrow$

   b. Write the equalities from Table 2 for each step in part (a).

   c. Write the conversion factors from the equalities in part (b).

   d. Perform the calculation, showing all work.

Exercises:

1. Write unit plans, and then perform the unit conversions below. Use only the conversion factors given in the problem or those given in Table 2.

   a. Convert 36 mg to g.

   complete the unit plan: mg $\rightarrow$ complete the equality: 1 mg =

   write the two conversion factors:

   perform the conversion:

   b. Convert 5.51 ms to μs.

   unit plan:

   conversion factors:

   perform the conversion:

2. A traveler wants to know the cost (in dollars) of the gasoline to travel 75 miles in the automobile in Model 1.

   a. Write a unit plan for the conversion.

   b. Write the two conversion factors needed for the conversion.

   c. Perform the calculation, showing all work and units.

3. What is the weight in grams of a 7.5-karat diamond engagement ring? There are exactly 200 mg in one karat. Make a unit plan first.

4. The speed of light is 299,792,458 m/s.
   a. What is this speed in km/hour? (Complete the unit plan: $\frac{m}{s} \rightarrow \frac{m}{min} \rightarrow \frac{km}{\quad} \rightarrow \frac{\quad}{\quad}$).

      Write the three equalities first.

   b. How long does it take for light to travel 100 meters? (Unit plan: m $\rightarrow$?)

Information: Units of volume

One *milliliter* (mL) is equal to one *cubic centimeter* (1 cc or 1 cm^3).

One *liter* (L) is equal to one *cubic decimeter* (1 dm^3).

5. Convert 1000 cm^3 to dm^3 using only the equalities given in the information above and the metric system prefix *milli* (10^{-3}). Show work. Make a unit plan first.

6. There are 5280 feet in a mile. Convert 161 km to miles, using only equalities taken from this activity. Make a unit plan first.

7. The daily dose of ampicillin for the treatment of an ear infection is 115 mg ampicillin per kilogram of body weight (115 mg/kg). What is the daily dose for a 27-lb child? Make a unit plan first.

8. Read the assigned pages in the text, and work the assigned problems.

Measurements and Significant Figures

(How do we measure things scientifically?)

Model 1: Outlines of a quarter and a glass bottle on top of a centimeter ruler

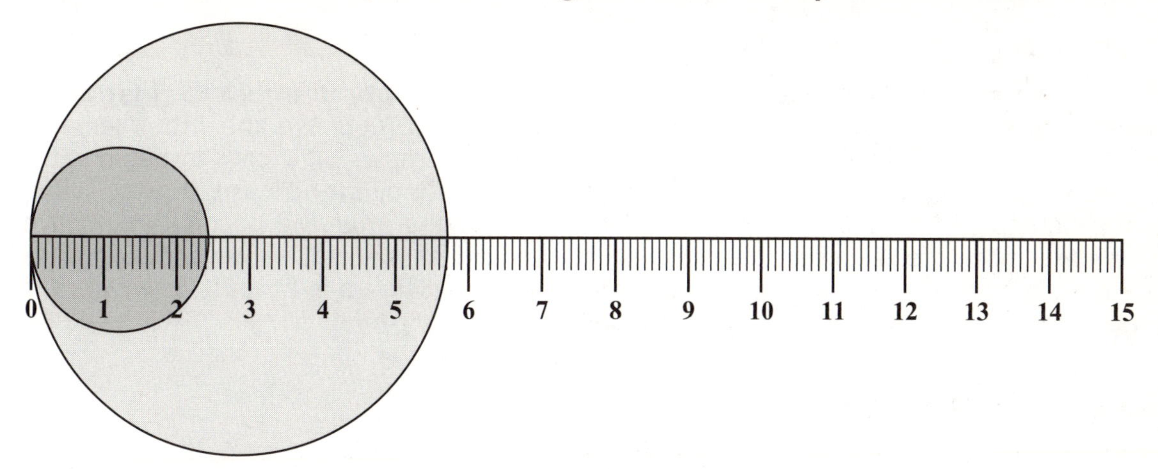

Critical Thinking Questions:

1. Use the ruler shown to determine the diameter of the quarter in Model 1 in centimeters. Express your answer as a **decimal number with 2 digits following the decimal point**. Include units.

2. Explain how you determined the final decimal place in the measurement in CTQ 1.

3. Use the ruler shown to determine the diameter of the glass bottle in Model 1 in centimeters. Express your answer as a decimal number with two digits after the decimal point. Include units.

4. Explain how you determined the final decimal place in the measurement in CTQ 3.

5. Write the two conversion factors that relate inches to centimeters.

6. Convert the diameter of the glass bottle to inches. Show work.

Information: Precision

Generally, when measurements are reported, there is **uncertainty in only the final digit**. This leads to a problem deciding how to report the answer to CTQ 6. We knew the diameter of the bottle to (at most) three digits of **precision** (for example, 5.70 cm) to begin with. These three digits are called **significant figures** ("sig figs") or **significant digits**. When the conversion to inches is performed, there are many digits after the decimal point. We cannot **create** more precision just by changing to the English system, so we should **round off** to three sig figs. If the digit following the third significant figure is at least 5, we should round **up**; otherwise, simply drop the trailing digits.

Critical Thinking Questions:

7. 2.244094488 cm rounded to three "sig figs" becomes __________________

8. 2.248031496 cm rounded to three "sig figs" becomes __________________

9. According to the U.S. mint, quarters are actually 2.426 cm in diameter. Does your measurement in CTQ 1 agree with this? Explain.

10. Convert the diameter of a quarter into inches. Show your work. Round your answer to the correct number of significant figures.

Information: Determining which figures in reported numbers are significant

Table 1: Rules for significant figures in measurements

Rule	Example (sig figs are underlined)
1. All non-zeroes are significant	$\underline{2}.\underline{25}$
2. Leading zeroes are **not** significant	$0.00\underline{54}$
3. Trapped zeroes **are** significant	$\underline{203}$, $0.0\underline{302}$
4. Trailing zeroes are significant **only** if an explicit decimal point is present*	$\underline{7}00$, $\underline{700}.$, $\underline{7}.\underline{00} \times 10^2$
5. Numbers that are **exact** by definition have an **infinite** number of significant figures. This includes conversions within a measurement system (and also 1 in = 2.54 cm).	$\underline{189}\ \text{cm} \times \frac{1\ \text{m}}{100\ \text{cm}} = \underline{1}.\underline{89}\ \text{m}$ (exact)

* See why scientific notation is useful? It eliminates any ambiguity.

Critical Thinking Questions:

11. Give the number of significant figures in each measurement.

    a. 100 m __________

    b. 0.00095 g ____________

    c. 3600.0 s __________

12. Write the following measurements in scientific notation, without changing the number of significant digits.

    a. 100 m __________

    b. 0.00095 g ____________

    c. 3600.0 s __________

13. Round each of the following to three significant digits.

    a. 0.00320700 L ______________

    b. 3.265×10^{-4} m ___________

    c. 129762 s ___________

Information: Significant figures in calculations

Table 2 gives some rules of thumb that help determine how much precision to report in the result of a calculation. These rules involve determining which digits in each number are **significant** (*i. e.,* they indicate the precision) and how to treat them in calculations.

These rules are simple to apply but only provide an **estimate** of the correct level of precision. This estimate is preferable to simply including all possible digits in the result of a calculation, but not as good as a rigorous error analysis.

Furthermore, since most calculations in chemistry involve multiplication or division, we can often use the following rule of thumb: Keep only as many significant figures in your answer as in the measurement with the fewest number of sig figs. Also, to avoid compounding errors, one should never round off intermediate results, but wait to round until you have the final answer.

Table 2: Rules for calculations with significant figures

Operation	Rule	Example (sig figs underlined)
Multiplication and division	Keep the smallest number of **sig figs.**	$\underline{25}0 \div \underline{7.134} = \underline{35}.043$ = **$\underline{35}$**
Addition and subtraction (do **after** multiplication and division)	Keep the smallest number of **decimal places.**	$\underline{73.147} + \underline{52.1} + 0.0\underline{5411} = \underline{125.3}0111$ = **125.3**
Logarithms	Only digits in the mantissa (after the decimal) are significant	$\log(\underline{2003}) = 3.\underline{3016}809$ = **3.$\underline{3017}$**

Exercises:

1. Consider the quarter in Model 1 and its diameter that you measured in centimeters.
   a. Write a unit plan to convert the diameter into millimeters.

   b. Write the two conversion factors needed.

   c. Perform the conversion, showing all work.

2. Follow the procedure in Exercise 1 to convert the diameter of the glass bottle into **micrometers**.

3. Perform the calculations, reporting your answer with the correct number of sig figs.
   a. 26.234 g – 5.6 g = ______________________
   b. 67.6 oz ÷ 8.0 oz/cup = ______________________cups
   c. $189 \text{ cm} + 6.0 \text{ in} \times \frac{2.54 \text{ cm}}{1 \text{ in}}$ = ______________________cm

4. Calculate the circumference of the glass bottle from Model 1 in centimeters. Report your answer with the correct number of significant figures. (Circumference = π × diameter.)

5. Consider the picture of the partially filled burette at the right. Circle the reading below that shows the correct level of precision. Explain your answer.

   15 mL   15.4 mL   15.40 mL   15.400 mL

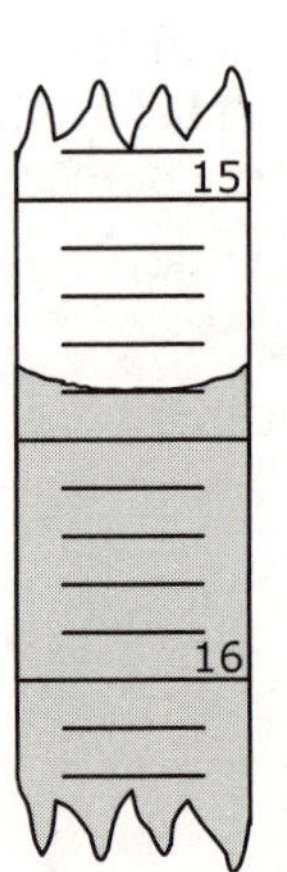

6. Read the assigned pages in your text, and work the assigned problems.

Density and Temperature

(What is measurement useful for?)

Critical Thinking Question:

1. Which weighs more, a ton of bricks or a ton of cotton balls?

Information: Density

CTQ 1 is a "trick" question that I remember as an early childhood "joke." Of course, the bricks and the cotton balls weigh the same (1 ton). But the cotton balls take up more space (volume), making them less **dense**. Density is a measure of the mass of a particular volume of something.

$$\text{density} = \frac{\text{mass}}{\text{volume}} \quad \textit{or} \quad d = \frac{m}{V}$$

Sometimes, something called **specific gravity** is used to report density. The specific gravity (sp gr) is a ratio of the density of something to the density of water.

$$\text{specific gravity (sp gr)} = \frac{\text{density of sample}}{\text{density of water}}$$

Since the density of water at room temperature is 1.00 g/mL, the sp gr of a sample is the same as the density, except that no units are given. If you know the sp gr, just add units of g/mL to give the density.

Suggested Demonstration: Density tower

Critical Thinking Questions:

2. A piece of tin metal weighing 85.251 g is placed into 41.1 cm^3 of ethyl alcohol (d = 0.798 g/cm^3) in a graduated cylinder, and it sinks to the bottom. The alcohol level increases to 52.8 cm^3.

   a. What is the mass of the tin?

   b. What is the volume of the tin?

   c. What is the density of the tin? Show work.

   d. What is the specific gravity of the tin?

3. If you have 750 mL of Everclear (95% ethyl alcohol, d = 0.80 g/mL), what mass of liquid do you have? (First, use algebra to rearrange the density equation to solve for *mass*.)

Model: Three temperature scales, Fahrenheit, Celsius and Kelvin

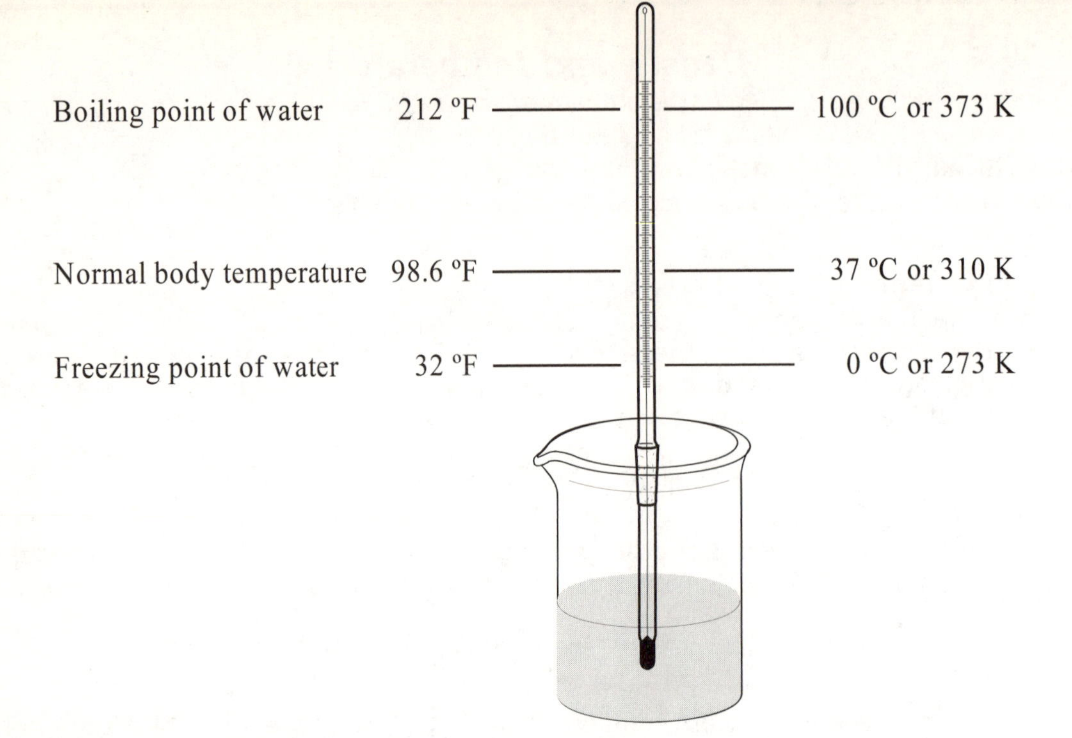

Normal body temperature is 98.6 degrees Fahrenheit (98.6 °F)
There are 66.6 Fahrenheit degrees (F°) between the freezing point of water and body temperature.

Critical Thinking Questions:

1. Write an equation to convert Celsius and Kelvins.

   K = °C + ________

2. How many Celsius degrees are there between the freezing and boiling points of water?_____
3. How many Kelvins are there between the freezing and boiling points of water? _____
4. Considering your answers to CTQs 3 and 4, how is the size of **one** Celsius degree related to the size of one Kelvin?
5. What is the boiling point of water in Fahrenheit? _______°F
6. How many Fahrenheit degrees are there between the freezing and boiling points of water?_____
7. Compare your answers to CTQs 2 and 6.
   a. Which has a larger size, a Celsius degree or a Fahrenheit degree?
   b. What is the **ratio** of the larger to the smaller degree? Include the units F° or C° where appropriate.

8. The freezing point of water is the reference (zero) point for the Celsius scale. What number of **Fahrenheit** degrees must be subtracted from the temperature at the freezing point of water so that it will also be the **zero point**?

9. So, to convert from °F to °C, two adjustments are needed. One has to subtract 32 to get to the same reference point (0° C), and then use a conversion factor (see CTQ 7). Using your answers to CTQ 7 and 8, write an equation for this. Do not look it up!

   °C =

10. To convert °C to °F, one has to use the conversion factor to convert to °F first, and then add 32 °F units ("adding apples to apples"). Write an equation for this.

   °F =

11. Suppose the room temperature is 72°F. Report this temperature in degrees Celsius and Kelvins.

Exercises:

1. The sp gr of mercury is 13.6. What is the volume of mercury in a barometer containing 2040 g of mercury?

2. Your long lost great aunt from France left you a fancy French oven in her will. Unfortunately, it is calibrated in Celsius, and you need to bake a cake at 350°F. At what temperature do you set the oven?

3. Some textbooks give the ratio 9/5 as the conversion factor between °F and °C. Explain how this is the same as the one in the equation you developed in CTQ 7.

4. Read the assigned pages in your text, and work the assigned problems.

*Electron Configuration and The Periodic Table**

(How are atoms and elements classified?)

Model 1: Diagrams of atoms the first three elements using the shell model

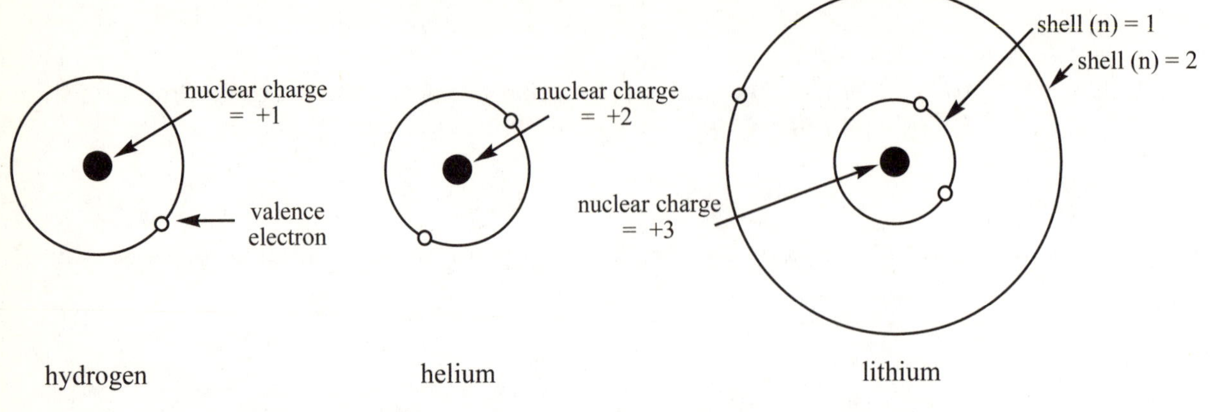

The outermost electron shell of an atom is its valence shell. Everything else is the core.

Critical Thinking Questions

1. How many electrons are in the valence shell of H? _____ of He? _____ of Li? _____
2. How many inner shell (core) electrons does H have? _____ He? _____ Li? _____
3. Considering everything in the core of the lithium atom, what is the net (total) charge in the core?
4. Explain how you arrived at your answer to CTQ 3.

Model 2: Diagrams of a lithium atom using the shell model (a) and the core charge concept (b)

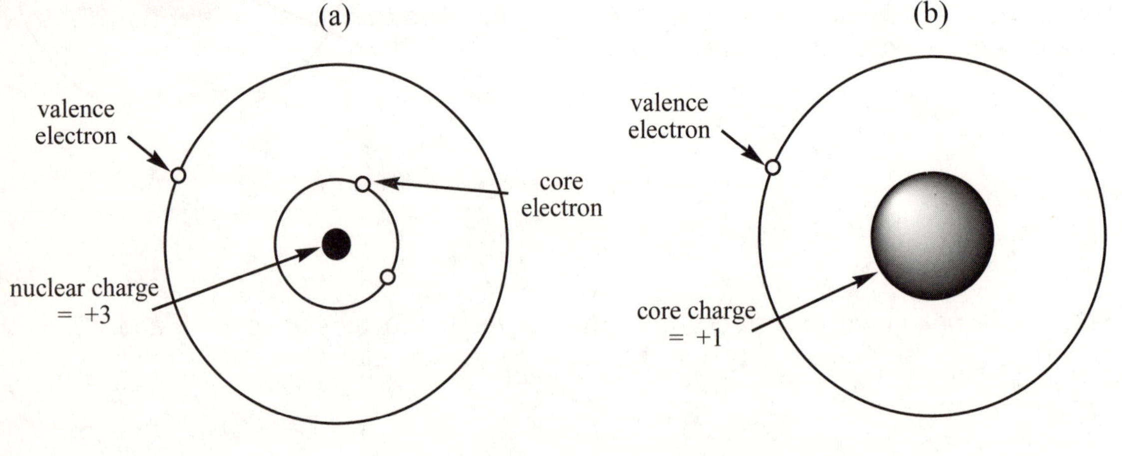

The valence electron "sees" a core charge of +1

* Adapted from ChemActivities 4 and 5, Moog, R.S.; Farrell, J.J. *Chemistry: A Guided Inquiry*, 3rd ed., Wiley, 2006, pp. 20-35.

Critical Thinking Questions:

Model 2 shows two ways of representing the third element, lithium. CTQs 5-8 will involve completing a similar diagram (below) for the fourth element, beryllium (Be).

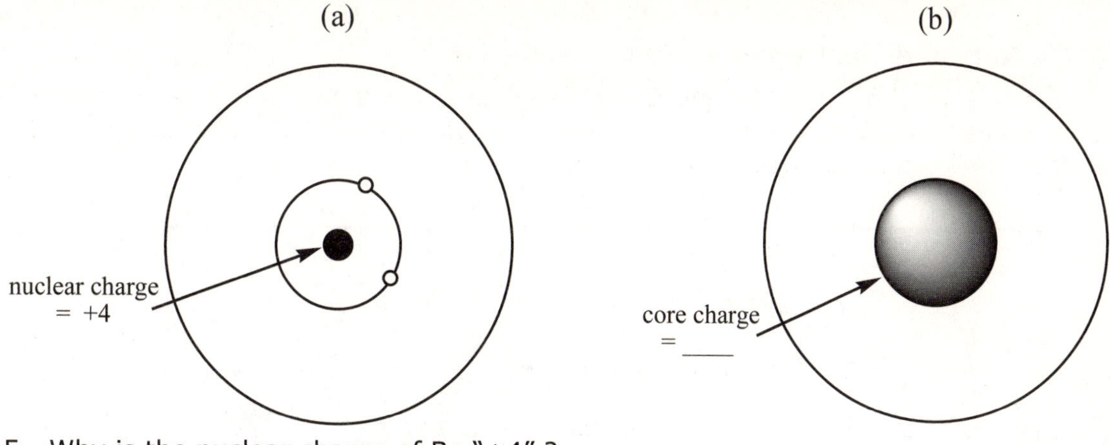

5. Why is the nuclear charge of Be "+4" ?

6. How many inner shell (core) electrons does Be have? _____

7. How many valence electrons would Be have? _____ Add them onto the shell model (a) in the diagram above.

8. Complete the diagram of a beryllium (Be) atom using the core charge concept (b).

Model 3: Diagrams of a sodium (Na) atom using the shell model (a) and the core charge concept (b)

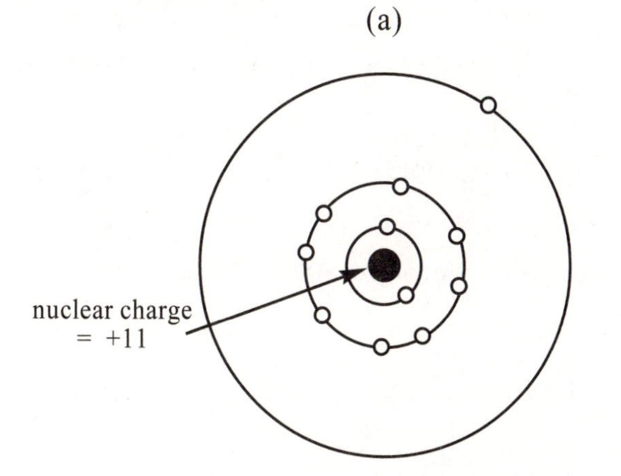

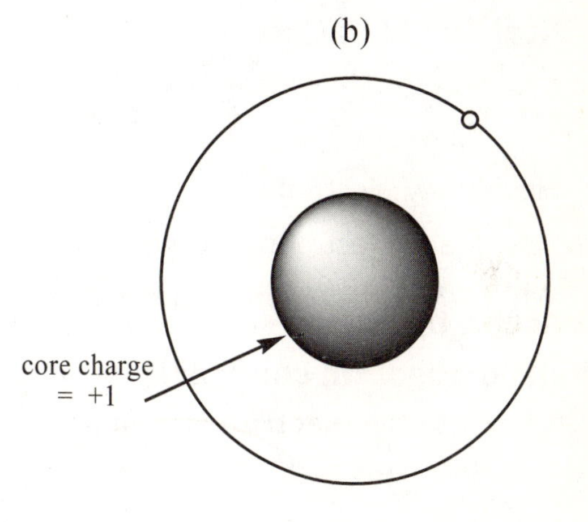

9. Label the shells n=1, n=2, and n=3 in Model 3(a).

10. How many electrons are in a full shell for n=1? _____

11. How many electrons are in a full shell for n=2? _____

12. How many valence electrons does Na have? _____

13. In one complete sentence, write a definition of the term *valence electron*.

14. What do H, Li and Na (and all group 1 elements) have in common?

Model 4: Orbitals and electron capacity in shells 1-4

Electron shells are not really two-dimensional circles like those shown in the earlier models. Electrons in shells are actually found in "cloud-like" regions of space called subshells, or **orbitals**. Each shell can contain one or more orbitals.

Shell (n)	Orbitals Available	Total Electron Capacity
1	one *s*	2
2	one *s*, three *p*	8
3	one *s*, three *p*, five *d*	18
4	one *s*, three *p*, five *d*, seven *f*	32

Model 5: Shapes of the first two types of orbitals, *s* and *p*

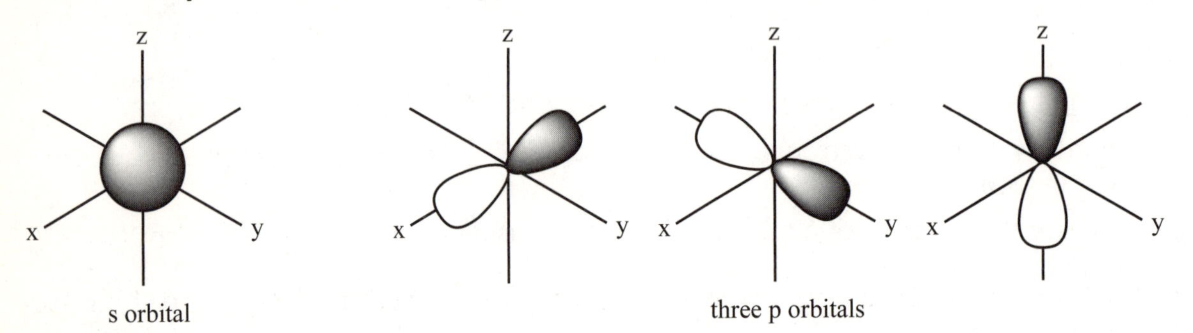

s orbital

three p orbitals

Critical Thinking Questions:

15. Explain how Model 4 shows that the three p orbitals in shell n = 2 can hold 6 electrons.

16. How many electrons can the five d orbitals in shell n=3 hold? _____

17. How many electrons can the f orbitals in shell n=4 hold? ______

18. What is the maximum number of electrons that can fit into any particular orbital? _____

Model 6: Electron configuration of sodium

We can write the **electron configuration** of Na like this:

$$1s^2\,2s^2\,2p^6 3s^1$$

So the electrons are really in clouds shaped like the pictures in Model 4. This is good to know, but most of the time, **we will stick to the simpler <u>shell</u> model, since it is easier to deal with.**

Critical Thinking Question:

19. Model 3 shows that the n=2 shell of the sodium atom contains 8 electrons. What information is given in Model 6 that is not present in Model 3? Give your answer in a complete sentence or two.

Model 7: Electron configuration of lead (optional)

After argon (atomic number 20), electrons do not fill neatly in order from shell $1 \rightarrow 2 \rightarrow 3$ *etc.* Consider, for example, the order that orbitals are filled to reach the electron configuration of the element lead (Pb), with 82 electrons:

$$1s^2\ 2s^2\ 2p^6\ 3s^2\ 3p^6\ 4s^2\ 3d^{10}\ 4p^6\ 5s^2\ 4d^{10}\ 5p^6\ 6s^2\ 4f^{14}\ 5d^{10}\ 6p^2$$

20. What is the largest shell (n) in lead that contains electrons? _____

21. How many valence electrons does lead have (see above)? _____ (Careful!)

22. Look at the periodic table attached to this activity (mostly blank, but with Pb shown). What is the relationship between the electron shell configuration of the element lead (Pb) and the row number on the left side of the table?

23. Count the width (number of columns) of each of the four "blocks." What is the relationship between the width and the maximum number of electrons per subshell? (Consider your answers to CTQs 13–16.)

Exercises:

1. Why are there only two elements in the first row of the periodic table?

2. Why does the second row not have a "d-block" section?

3. Find the element "A" on the blank periodic table.

   a) Based on its position in the s-block, what would be the **last** entry in its electron configuration? _____ (Do this without counting electrons.)

   b) How many valence electrons does element "A" have? ____

4. Find the element "B" on the blank periodic table.

   a) Based on its position in the p-block, what would be the **last** entry in its electron configuration? _____ Do this without counting electrons.

   b) How many valence electrons does element "B" have?____

(Caution: The procedure followed in Exercises 3 and 4 becomes more complicated for elements in the *d* and *f*-blocks. Read your text if you are interested in the relationship between the periodic table and the electron configuration.)

5. A student makes the following statement: "The electron configuration of all halogens ends in np^5." Is the student correct? Explain.

6. Read the assigned pages in your text, and work the assigned problems.

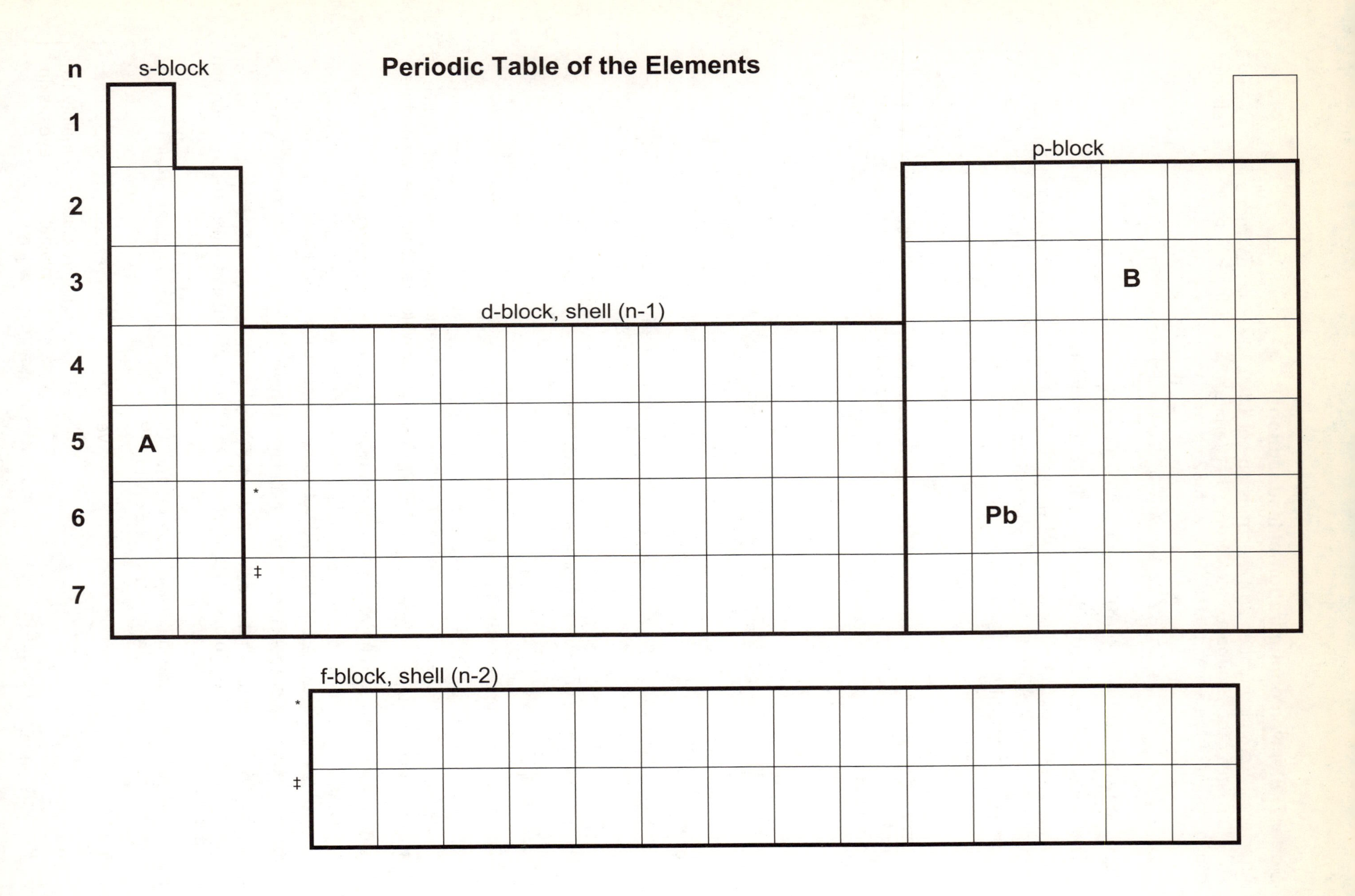
Periodic Table of the Elements
n
1
2
3
4
5
6
7
s-block
d-block, shell (n-1)
p-block
f-block, shell (n-2)
A
B
Pb
*
‡

Nuclear Chemistry

(What is radiation?)

Model: Nuclide symbols for three isotopes of carbon

$^{12}_{6}C$	$^{13}_{6}C$	$^{14}_{6}C$
carbon-12	carbon-13	carbon-14

Critical Thinking Questions:

1. How many protons are in carbon-12? ____ How many neutrons are in carbon-12? ____
2. How many protons are in carbon-13? ____ How many neutrons are in carbon-13? ____
3. How many protons are in carbon-14? ____ How many neutrons are in carbon-14? ____
4. Make a list of what is the same and what is different among isotopes.

5. What does the subscripted 6 represent in all three nuclide symbols in the Model?

Information: Nuclear reactions and ionizing radiation

A **nuclear reaction** is a change in the composition of the nucleus of an atom. This is not normally considered a chemical reaction, and does not depend on what molecule the atom might be in.

There are three types of nuclear reactions: fusion, fission, and radioactivity. Fusion (combining of nuclei into larger nuclei, such as in stars and the sun) and fission ("splitting the atom," such as in a nuclear reactor) do not concern us much in chemistry.

Some isotopes are radioactive, meaning that their nuclei break down ("decay") and give off particles, "rays," or both. There is no simple way to predict which isotopes are radioactive.

Table 1: Some types of ionizing radiation produced in nuclear reactions

Type of Radiation	Symbol	Mass Number	Charge	Relative penetrating ability	Shielding required	Biological hazard
Alpha particle	α, $^{4}_{2}He$	4	2+	very low	clothing	none unless inhaled
Beta particle	β, $^{0}_{-1}e$	0	1−	low	heavy cloth, plastic	mainly to eyes, skin
Gamma ray	γ, $^{0}_{0}\gamma$	0	0	very high	lead or concrete	whole body
Neutron	$^{1}_{0}n$	1	0	very high	water, lead	whole body
Positron	β^{+}, $^{0}_{1}e$	0	1+	low	heavy cloth, plastic	mainly to eyes, skin

Critical Thinking Questions:

6. What does the subscript indicate in the **symbols** in Table 1?

7. Explain how your answer to CTQ 6 is consistent with your answer to CTQ 5.

8. Consider the following nuclear reaction: $^{238}_{92}U \rightarrow ^{234}_{90}Th + ^{4}_{2}He$

   a. What type of radioactivity is produced?

   b. How does the number of protons in the reactant compare with the total number of protons in the products?

   c. How does the number of neutrons in the reactant compare with the total number of neutrons in the products?

   d. How does the mass number of the reactant compare with the total of the mass numbers of the products?

   e. Show how each side of the reaction equation would change if a gamma ray were also released in the process.

9. Balance the mass numbers and "atomic numbers" to complete the equation.

   a. $^{131}_{53}I \rightarrow ^{0}_{-1}e \quad + \quad \square$

   b. What type of radioactivity is given off in this reaction?

Table 2: Half-lives of some radioisotopes

Radioisotope	Radiation type	Half-life	Use
barium-131	γ	11.6 days	detection of bone tumors
carbon-14	β	5730 yr	carbon dating
chromium-51	γ, X-rays	27.8 days	measuring blood volume
cobalt-60	β, γ	5.3 yr	food irradiation, cancer therapy
iodine-131	β	8.1 days	hyperthyroid treatment
uranium-238	α, β, γ	4.47×10^9 yr	dating igneous rocks

The time required for half of a sample of a radioactive isotopes to decay is called the half-life ($t_{½}$).

Critical Thinking Questions:

10. Consider a 100-gram sample of radioactive cobalt-60.
    a. How much time will it take before half the sample has decayed?

    b. Approximately how many grams of radioactive cobalt-60 will remain after 11 years?

11. Consider a sample of iodine-131.
    a. How many half-lives would it take for the sample to decay until less than 1% of the original isotope remained?

    b. How many days would this be?

12. Considering only the half lives of uranium-238 and iodine-131, which would be more appropriate for internal usage (ingestion) for medical tests? Explain.

Exercises:

1. After an organism dies, it stops taking in radioactive carbon-14 from the environment. If the carbon-14:carbon-12 ratio ($^{14}_{6}C/^{12}_{6}C$) in a piece of petrified wood is one sixteenth of the ratio in living matter, how old is the rock? (Hint: How many half lives have elapsed?)

2. Would chromium-51 be useful for dating rocks containing chromium? Why or why not?

3. Suppose that 0.50 grams of barium-131 are administered orally to a patient. Approximately how many milligrams of the barium would still be radioactive two months later?

4. Complete the equations.

   a. $^{30}_{15}P \longrightarrow ^{0}_{1}e +$ (What type of radiation is this? ________________)

   b. $^{113}_{47}Ag \xrightarrow{\text{beta decay}}$

   c. $\xrightarrow{\alpha \text{ and } \gamma \text{ emission}} ^{222}_{86}Rn + \quad + ^{0}_{0}\gamma$

5. Read the assigned pages in your textbook and work the assigned problems.

Ions and Ionic Compounds

(What are ionic compounds?)

Model 1: Common charges (by group) on elements *when in ionic compounds*.

Grp # 1	2	3	4	5	6	7	8	9	10	11	12	13	14	15	16	17	18
Chg: +1	+2	← varies →										+3		-3	-2	-1	
1 H																	2 He
3 Li	4 Be											5 B	6 C	7 N	8 O	9 F	10 Ne
11 Na	12 Mg											13 Al	14 Si	15 P	16 S	17 Cl	18 Ar
19 K	20 Ca	21 Sc	22 Ti	23 V	24 Cr	25 Mn	26 Fe	27 Co	28 Ni	29 Cu	30 Zn	31 Ga	32 Ge	33 As	34 Se	35 Br	36 Kr
37 Rb	38 Sr	39 Y	40 Zr	41 Nb	42 Mo	43 Tc	44 Ru	45 Rh	46 Pd	47 Ag	48 Cd	49 In	50 Sn	51 Sb	52 Te	53 I	54 Xe
55 Cs	56 Ba	57 La	72 Hf	73 Ta	74 W	75 Re	76 Os	77 Ir	78 Pt	79 Au	80 Hg	81 Tl	82 Pb	83 Bi	84 Po	85 At	86 Rn
87 Fr	88 Ra	89 Ac	104 Rf	105 Db	106 Sg	107 Bh	108 Hs	109 Mt	110 Ds	111	112		114		116		

A cation has a positive charge. An anion has a negative charge.

Critical Thinking Questions:

1. In an ionic compound, what is the charge (sign and magnitude) on an ion from

   a. group 1? ______

   b. group 2?______

   c. group 13?______

2. Do metals typically form **anions** or **cations** (circle one)?

3. In an ionic compound, what is the charge on an ion from

   a. group 15? ______

   b. group 16?______

   c. group 17?______

4. Do nonmetals typically form **anions** or **cations** (circle one)?

Model 2: Formulas and names of some binary ionic compounds

Formula	Group Number of Metal	Name
NaBr	1	sodium bromide
K_2O	1	potassium oxide
MgF_2	2	magnesium fluoride
FeS	8	iron(II) sulfide
Fe_2S_3	8	iron(III) sulfide
Cu_2S	11	copper(I) sulfide
InP	13	indium phosphide
PbI_2	14	lead(II) iodide

Critical Thinking Questions:

5. In the name of an ionic compound, which ion is always written first— **the anion** or **the cation** (circle one).

6. Circle the **names** of compounds in Model 2 for which the metal ion is **not** in groups 1, 2, or 13. How does the name of the *cation* differ from the name of the *metal*?

7. For metal ions from groups 1, 2, or 13 in ionic compounds, how does the name of the cation differ from the name of the metal?

8. For nonmetals, how does the name of the anion differ from the name of the element?

9. Consider the formula for magnesium fluoride, MgF_2.
   a. What is the charge on the magnesium ion? _____
   b. What is the charge on the fluoride ion? _____
   c. What is the overall (total) charge on the compound? _____

10. What is the overall charge on the compound sodium bromide (NaBr)? _____

11. In general, what is the overall charge on an ionic compound? _____

12. What is the charge on iron in FeS?_____ In Fe_2S_3? _____

13. Based on your answer to CTQ 12, what do the Roman numerals in Model 2 represent?

14. Complete the rules for naming ionic compounds:
    - Naming metal ions: name the metal [example: Ca^{2+} = ____________________]
    - If the metal is **not** in group 1, 2, or 13, add a Roman numeral in parentheses that represents ________________________________[*e. g.*, Fe^{3+} =____________]
    - Nonmetals: change ending of element name to ________ [*e. g.*, N^{3-} = ____________]
    - Naming ionic compounds: name the cation, then the anion [example: FeN = ________________________________]

Table 1: Formulas and names of some common polyatomic ions to memorize

Formula	Name	Formula	Name
NH_4^+	ammonium	MnO_4^-	permanganate
$C_2H_3O_2^-$	acetate	CO_3^{2-}	carbonate
CN^-	cyanide	$Cr_2O_7^{2-}$	dichromate
OH^-	hydroxide	SO_4^{2-}	sulfate
OCl^-	hypochlorite	PO_4^{3-}	phosphate
NO_3^-	nitrate		

Table 2: Rules for naming other polyatomic ions

Rule	Examples
1. If adding a H^+ to a polyatomic ion results in a new ion, add the word "hydrogen" in front of the name; if adding 2 H^+, add the word "dihydrogen."	HPO_4^{2-} = hydrogen phosphate $H_2PO_4^-$ = dihydrogen phosphate
2. For ions with one fewer oxygen atoms than the common ion, change the ending from "-ate" to "-ite"	SO_4^{2-} = sulf**ate**, so SO_3^{2-} = sulf**ite**

Exercises:

1. Complete the following table.

Formula	Name
$CoCl_4$	__________
__________	potassium nitrate
$Ba(OH)_2$	__________
__________	sodium hydrogen carbonate
__________	beryllium bromide
Li_2CO_3	__________
__________	copper(II) oxide
__________	sodium hypochlorite
$Ca(C_2H_3O_2)_2$	__________
__________	magnesium sulfate
$NaNO_2$	__________
__________	vanadium(III) sulfate

2. Based on their electron shell configurations, give a rationalization for why all alkaline earth metals in ionic compounds have a +2 charge.

3. Based on their electron shell configurations, give a rationalization for why all halogens in ionic compounds have a −1 charge.

4. Read the assigned pages in the text, and work the assigned problems.

Covalent and Ionic Bonds

(Why do atoms bond together?)

Model 1: Two types of chemical bonding*

Ions held together by opposite charges are said to be ionically bonded.
Ionic compounds contain ions—typically a metal ion along with nonmetals.

Atoms sharing valence electrons are said to be covalently bonded.
Covalent molecules typically contain only nonmetals.

Model 2: Lewis electron-dot structures for hydrogen and the second row elements

•H •Li •Be •B• •C• •N• :O• :F: :Ne:

Critical Thinking Questions:

1. Which two elements in Model 2 are metals?
2. Are these two elements likely to be in a covalent molecule? Explain.
3. Consider Model 2. How is the number of dots related to the number of valence electrons?
4. By extension, write the electron-dot (Lewis) structures for sulfur, chlorine, and sodium.
5. The ions formed in compounds from group 1 atoms (the alkali metals) are almost always M^+ ions (that is, they have a +1 charge). Considering Model 2, explain this result.
6. The ions formed in compounds from group 2 atoms (the alkaline earth metals) are almost always M^{2+} ions. Explain this result.

* A third type of bonding, metallic bonding, will not be considered in this book.

7. The ions formed in compounds from group 17 atoms (the halogens) are almost always X^- ions. Explain this result.

8. Write electron dot (Lewis) structures for each of the **ions** F^-, Cl^-, and Br^-.

Model 3: Typical number of covalent bonds for elements common in biology

Element	Number of Bonds
H	1
O	2
N	3
C	4

Critical Thinking Questions:

9. How many **electrons** in the dot structure for O in Model 2 are **paired**? ____
10. How many **unpaired** electrons are in the dot structure for O? ____
11. How many electrons in the dot structure for N are paired? ____
12. How many unpaired electrons are in the dot structure for N? ____
13. How many electrons in the dot structure for C are paired? ____
14. How many unpaired electrons are in the dot structure for C? ____
15. How is the number of covalent bonds that an atom makes related to its electron-dot structure?

16. Which nonmetal in Model 2 is unlikely to be in a covalent molecule? Explain.

Model 4: Covalent bonding (sharing valence electrons) between H and F

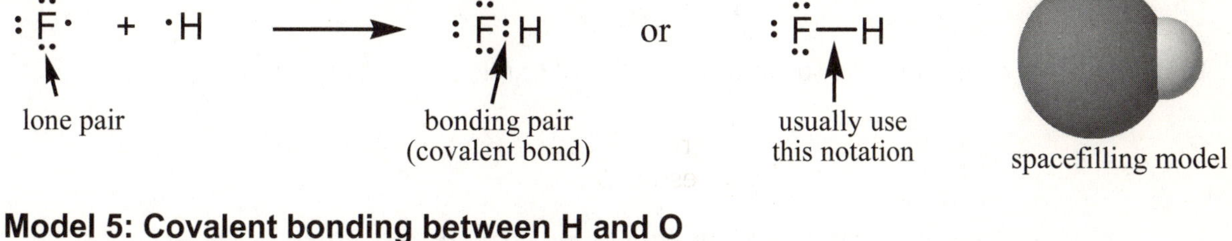

Model 5: Covalent bonding between H and O

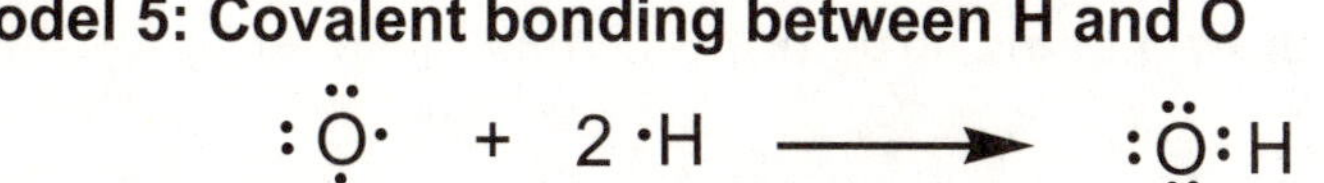

Critical Thinking Questions:

17. Given the shell model of the atom, why do you think that Lewis proposed a maximum of two electrons for hydrogen and eight for carbon, nitrogen, oxygen, and fluorine atoms?

18. Answer the following for the phosphorus atom:

    a. How many valence electrons does P have? _____

    b. What is the Lewis representation for P?

    c. How many additional electrons does P need when it forms a molecule? _____

    d. What is the likely formula for a molecule composed of hydrogen atoms and one phosphorus atom? _____ Draw the Lewis structure.

19. Answer the following for the sulfur atom:

    a. How many valence electrons does S have? _____

    b. What is the Lewis representation for S?

    c. How many additional electrons does S need when it forms a molecule? _____

    d. What is the likely formula for a molecule composed of hydrogen atoms and one sulfur atom? _____ Would this compound be **ionic** or **molecular** (circle one)? Explain.

    e. Draw the Lewis structure.

20. a. What is the likely formula for a molecule composed of sodium atoms and one sulfur atom?

    b. Would this compound be **ionic** or **molecular** (circle one)? Explain.

    c. Why would it be inappropriate to draw a Lewis structure (such as in Model 5) for this compound?

21. Covalent compounds form *individual units* called **molecules**. Compare the model of the HF molecule in Model 4 with the model of a solid sodium chloride (NaCl) crystal at the right [dotted lines shown to enhance the three-dimensionality of the crystal]. Then, explain why you think the word "molecule" is not used for ionic compounds.

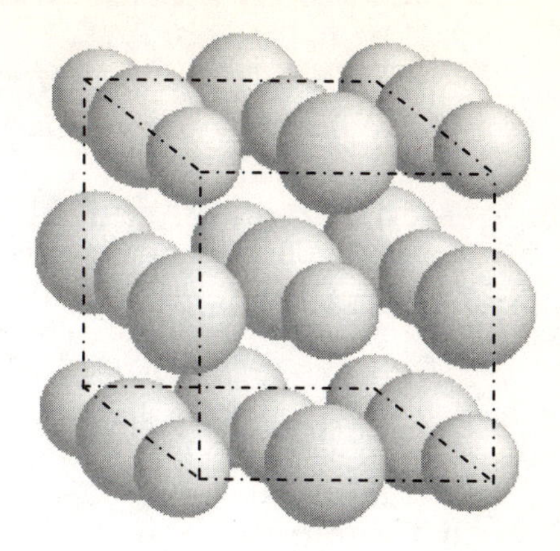

Exercises:

1. Draw reasonable Lewis (electron-dot) structures for the following molecules:
   a. NH_3 (ammonia)

   b. NCl_3 (nitrogen trichloride)

   c. CH_4 (methane)

   d. $CHCl_3$ (chloroform)

2. In NaCl, what are the charges on the ions? Explain your reasoning in terms of the Lewis (electron-dot) representations of Na and Cl.

3. In CTQ 21, the two different spheres are different sizes. Which spheres represent the sodium ions and which represent the chloride ions? Hint: Consider the electron shell configurations of the two ions.

4. Read the assigned pages in the text, and work the assigned problems.

Electrolytes, Acids, and Bases

(Which compounds produce ions when dissolved in water?)

Suggested Demonstration: Electrolytes

Model 1: Electrolytes

Only separate, charged particles (such as ions) can carry electrical currents.

Electrolytes can carry an electrical current when dissolved in water.

Critical Thinking Question:

1. What happens to electrolytes when they dissolve in water?

Model 2: Types of electrolytes

In water solution, *strong electrolytes* dissociate completely into ions, *weak electrolytes* dissociate only slightly, and *nonelectrolytes* dissociate undetectably or not at all.

Critical Thinking Question:

2. Describe a method by which you could tell if a particular solution contains a strong electrolyte, weak electrolyte, or nonelectrolyte.

Model 3: Some common acids and bases

Type of electrolyte	Acids	Bases
Strong (any acid or base not listed here is weak)	HCl	LiOH
	HBr	NaOH
	HI	KOH
	H_2SO_4	$Ca(OH)_2$
	HNO_3	$Sr(OH)_2$
	$HClO_4$	$Ba(OH)_2$
Weak	$HC_2H_3O_2$	$Mg(OH)_2$
	HCN	

Acids dissociate in water to give hydrogen (H^+) ions and an anion. Bases dissociate in water to give hydroxide (OH^-) ions and a cation. *Strong* acids and bases dissociate *completely*. Note that the acids are molecules, while the bases are ionic compounds.

Critical Thinking Questions:

3. Consider Model 3. What do all the molecular formulas of the acids have in common?

4. How can you recognize an acid from the molecular formula?

5. How can you recognize a base from its formula?

6. What happens to strong acids when dissolved in water?

7. What happens to weak acids when dissolved in water?

8. Are all acids strong electrolytes? Explain.

9. Describe what happens to the ions in solid sodium hydroxide (NaOH) during the process of dissolving in water.

Model 4: Solubilities of some ionic compounds

Ionic compound	Solubility in water	Type of electrolyte	Major species present when dissolved in water
$MgCl_2$	soluble	strong	$Mg^{2+}(aq)$, $Cl^{-}(aq)$
MgO	insoluble	nonelectrolyte	MgO(s)
K_2S	soluble	strong	$K^{+}(aq)$, $S^{2-}(aq)$
CuS	insoluble	nonelectrolyte	CuS(s)
$Ca(NO_3)_2$	soluble	strong	$Ca^{2+}(aq)$, $NO_3^{-}(aq)$
$Ca_3(PO_4)_2$	insoluble	nonelectrolyte	$Ca_3(PO_4)_2(s)$

Molecular compounds other than acids and bases, and *insoluble ionic compounds* do not dissociate in water and so are nonelectrolytes.

The phase label *(aq)*, meaning "aqueous," can be used to show that a species is dissolved in water. **We will not consider how to predict if an ionic compound is soluble in water until later in the course.**

Critical Thinking Questions:

10. A solution of $MgCl_2$ in water could be written as $MgCl_2(aq)$. Besides water, what species would actually be present in the solution?

11. MgO is an ionic compound that does not dissolve in water, it would just collect as a solid—*i. e.,* MgO(s)—at the bottom of the container. Would there be any ions dissolved in the water? Explain.

12. Describe a method by which you could tell if an ionic compound is soluble in water.

13. Consider Model 4. How is the solubility of an ionic compound related to its classification as an electrolyte?

14. The following compounds from Model 4 are placed in water. Add the phase labels (s) or (aq) to represent whether the compounds are soluble or not:

K_2S CuS $Ca(NO_3)_2$ $Ca_3(PO_4)_2$

15. Suppose that each compound in the table below is placed in water. The phase labels given describe whether the compound dissolves or not. Fill in the table with the properties for each compound.

Compound	**Acid**, other **molecule, base,** or other **ionic compound**	**Strong**, **weak**, or **non**electrolyte
HBr(aq)		
KOH(aq)		
$CH_2O(aq)$		
$Ca_3(PO_4)_2(s)$		
$Al(OH)_3(s)$		
HOCl(aq)		
$H_2S(aq)$		
$Fe_2O_3(s)$		
$Na_2S(aq)$		

Exercises:

1. Write formulas for the major species present in the solutions from CTQ 15.

Compound	Major species present when dissolved in water (or "need more information")
$HBr(aq)$	
$KOH(aq)$	
$CH_2O(aq)$	
$Ca_3(PO_4)_2(s)$	
$Al(OH)_3(s)$	
$HOCl(aq)$	
$H_2S(aq)$	
$Fe_2O_3(s)$	
$Na_2S(aq)$	

2. Read the assigned pages in your text, and work the assigned problems.

Naming Binary Molecules, Acids, and Bases

(How are chemical formulas and names related?)

Information: Naming Binary Molecules

Recall that molecules (as opposed to ionic compounds) are formed from nonmetals only. There is a simple method to name binary molecules (molecules containing only two elements).

1. Name the first element, using a prefix to indicate the number of atoms in the formula.
2. Name the second element, using a prefix to indicate the number of atoms in the formula, and changing the ending to "-ide."
3. Remove the vowel at the end of the prefix if it seems awkward.
4. Do not use the prefix "mono-" on the first element.

Table 1: Prefixes to indicate the number of atoms of an element in a binary molecule

Number	Prefix	Number	Prefix
1	mono-	6	hexa-
2	di-	7	hepta-
3	tri-	8	octa-
4	tetra-	9	nona-
5	penta-	10	deca-

Examples: NI_3 = nitrogen triiodide; N_2O = dinitrogen monoxide (not "mon**oo**xide"); N_2O_5 = dinitrogen pentoxide (not "pent**ao**xide")

Critical Thinking Questions:

1. The following is from "Ban Dihydrogen Monoxide," Coalition to Ban DHMO, 1988. Quoted in http://www.dhmo.org/truth/Dihydrogen-Monoxide.html [accessed 03 Feb 2006]:

   Dihydrogen monoxide is colorless, odorless, tasteless, and kills uncounted thousands of people every year. Most of these deaths are caused by accidental inhalation of DHMO, but the dangers of dihydrogen monoxide do not end there. Prolonged exposure to its solid form causes severe tissue damage. Symptoms of DHMO ingestion can include excessive sweating and urination, and possibly a bloated feeling, nausea, vomiting and body electrolyte imbalance. For those who have become dependent, DHMO withdrawal means certain death.

   Explain how these claims are true or false.

2. Explain what is wrong with the following molecule names. Then give the correct name.

   a. NO nitrogen oxide

   b. CO monocarbon monoxide

   c. CS_2 carbon disulfate

Information: Naming the Binary Acids

When the binary acids hydrogen chloride (HCl), hydrogen bromide (HBr), and hydrogen iodide (HI) are dissolved in water (aq), they are called "hydrochloric acid," "hydrobromic acid," and "hydroiodic acid" respectively.

Critical Thinking Questions:

3. Based on the names of HCl, HBr, and HI, what would HF(aq) be called?

4. Is HF a **strong** or **weak** acid (circle one)? Explain.

Information: Naming Oxoacids

An **oxoacid** is an acid that contains hydrogen, oxygen, and one other element. Oxoacids may be formed by adding H^+ to polyatomic ions. One H^+ is added for each negative charge on the ion (example: the charge on sulfate is −2, so add two H^+ ions to make H_2SO_4).

A. if the ion has an ending of "-ate," the acid has an ending of "-ic" (example: SO_4^{2-} = "sulf**ate**," so H_2SO_4 = "sulfur**ic** acid").
   - memory device: "**Ic**k," I "**ate**" the acid!

B. if the ion has an ending of "-ite," the acid has an ending of "-ous" (example: SO_3^{2-} = "sulf**ite**," so H_2SO_3 = "sulfur**ous** acid").

Critical Thinking Questions:

5. Name the following oxoacids. (You may wish to consult the polyatomic ion names in ChemActivity 9, Table 1.)

   a. HOCl ________________________

   b. HNO_3 ________________________

   c. HNO_2 ________________________

   d. $HC_2H_3O_2$ ________________________

Information: Summary flowchart for naming binary ionic and molecular compounds

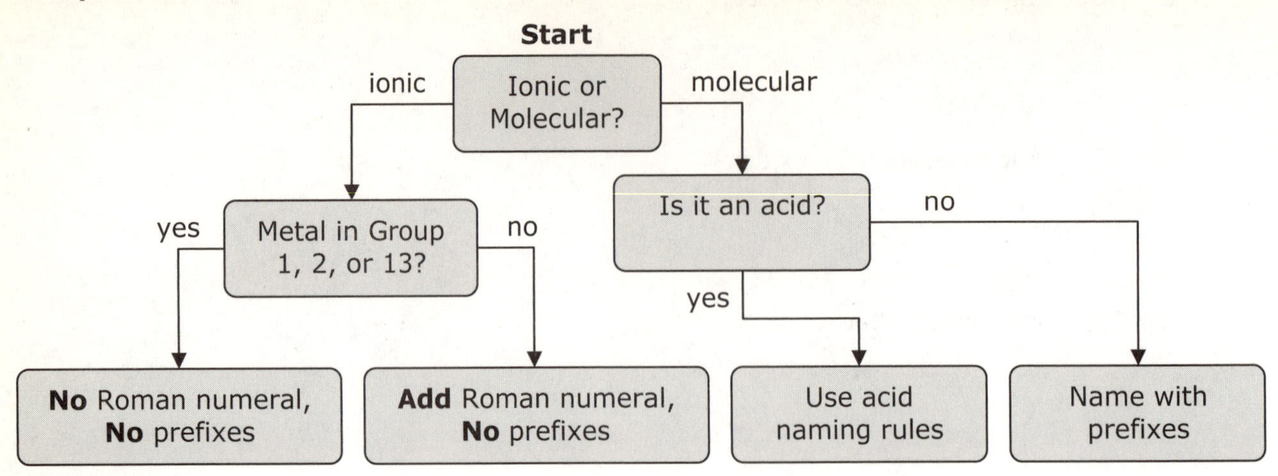

Exercises:

1. Fill in the table with the name of each compound.

Compound	Name
MnS	
CS_2	
Na_2S	
$MgSO_4$	
H_2SO_3	
H_3PO_4	
CuOH	
P_2O_5	

2. Fill in the table with the formula of each compound.

Compound	Name
	acetic acid
	dinitrogen monoxide
	copper(I) oxide
	carbonic acid
	magnesium bromide
	oxygen difluoride
	sodium nitrite
	tin(II) hydroxide

3. Read the assigned pages in your text, and work the assigned problems.

Molecular Shapes[*]

(What shapes do molecules have?)

Model 1: Bond angle and electron fomains.

A **bond angle** is the angle made by three connected nuclei in a molecule. By convention, the bond angle is considered to be between 0° and 180°.

Table 1: Bond angles and electron domains in selected molecules.[†]

Molecular Formula	Lewis Structure	Bond Angle (CAChe)	No. of Bonding Domains (central atom)	No. of Nonbonding Domains (central atom)
CO_2	:Ö=C=Ö:	∠OCO = 180°	2	0
HCCH	H–C≡C–H	∠HCC = 180°	2	0
H_2CCCH_2	H–C(H)=C=C(H)–H	∠CCC = 180°	2	0
ClNNCl	:C̈l–N̈=N̈–C̈l:	∠ClNN = 117.4°	2	1
NO_3^-	⊖:Ö–N⊕(=Ö:)–Ö:⊖	∠ONO = 120°	3	0
H_2CCH_2	H–C(H)=C(H)–H	∠HCH = 121.1°	3	0

(Table 1 continues on the next page)

[*] Adapted from ChemActivity 18, Moog, R.S.; Farrell, J.J. *Chemistry: A Guided Inquiry*, 3rd ed., pp. 102–110, Copyright 2006, John Wiley & Sons, Inc. Models, Tables, Figures, and Critical Thinking Questions reprinted with permission of John Wiley & Sons, Inc.

[†] Bond angles calculated with MOPAC (Oxford Molecular, CAChe). MOPAC calculations yield bond orders, bond lengths, and bond angles that are generally in good agreement with experimental evidence.

Table 1 (continued): Bond angles and electron domains in selected molecules.

Molecular Formula	Lewis Structure	Bond Angle (CAChe)	No. of Bonding Domains (central atom)	No. of Nonbonding Domains (central atom)
CH_4	H | H–C–H | H	∠HCH = 109.45	4	0
CH_3F	:F̤̈: | H–C–H | H	∠HCH = 109.45° ∠HCF = 109.45°	4	0
CH_3Cl	:C̈l: | H–C–H | H	∠HCH = 109.45° ∠HCCl = 109.45°	4	0
CCl_4	:C̈l: | :C̤̈l–C–C̤̈l: | :C̤l:	∠ClCCl = 109.45°	4	0
NH_3	H–N̈–H | H	∠HNH = 107°	3	1
NH_2F	H–N̈–F̤̈: | H	∠HNH =106.95° ∠HNF = 106.46°	3	1
H_2O	:Ö–H | H	∠HOH = 104.5°	2	2

Critical Thinking Questions:

1. How is the number of bonding domains on a given atom within a molecule (such as those in Table 1) determined?

2. How is the number of nonbonding domains on a given atom within a molecule (such as those in Table 1) determined?

3. The bond angles in Table 1 can be grouped, roughly, around three values. What are these three values?

4. What correlation can be made between the values in the last two columns in Table 1 and the groupings identified in CTQ 3?

Model 2: Models for methane, ammonia, and water.

Use a molecular modeling set to make the following molecules: CH_4; NH_3; H_2O. In many model kits: carbon is black; oxygen is red; nitrogen is blue; hydrogen is white; use the short, gray links for single bonds. Nonbonding electrons are not represented in these models.

Critical Thinking Questions:

5. Sketch a picture of the following molecules based on your models: CH_4; NH_3; H_2O.

6. Describe (with a word or short phrase) the shape of each of these molecules: CH_4; NH_3; H_2O.

Model 3: Types of electron domains.

A domain of electrons (two electrons in a **nonbonding domain**, sometimes called a **lone pair**; two electrons in a **single bond domain**; four electrons in a **double bond domain**; six electrons in a **triple bond domain**) tends to repel other domains of electrons. Domains of electrons around a central atom will orient themselves to minimize the electron-electron repulsion between the domains.

Figure 1. Minimization of electron-electron repulsion leads to a unique geometry for two, three, and four domains of electrons.

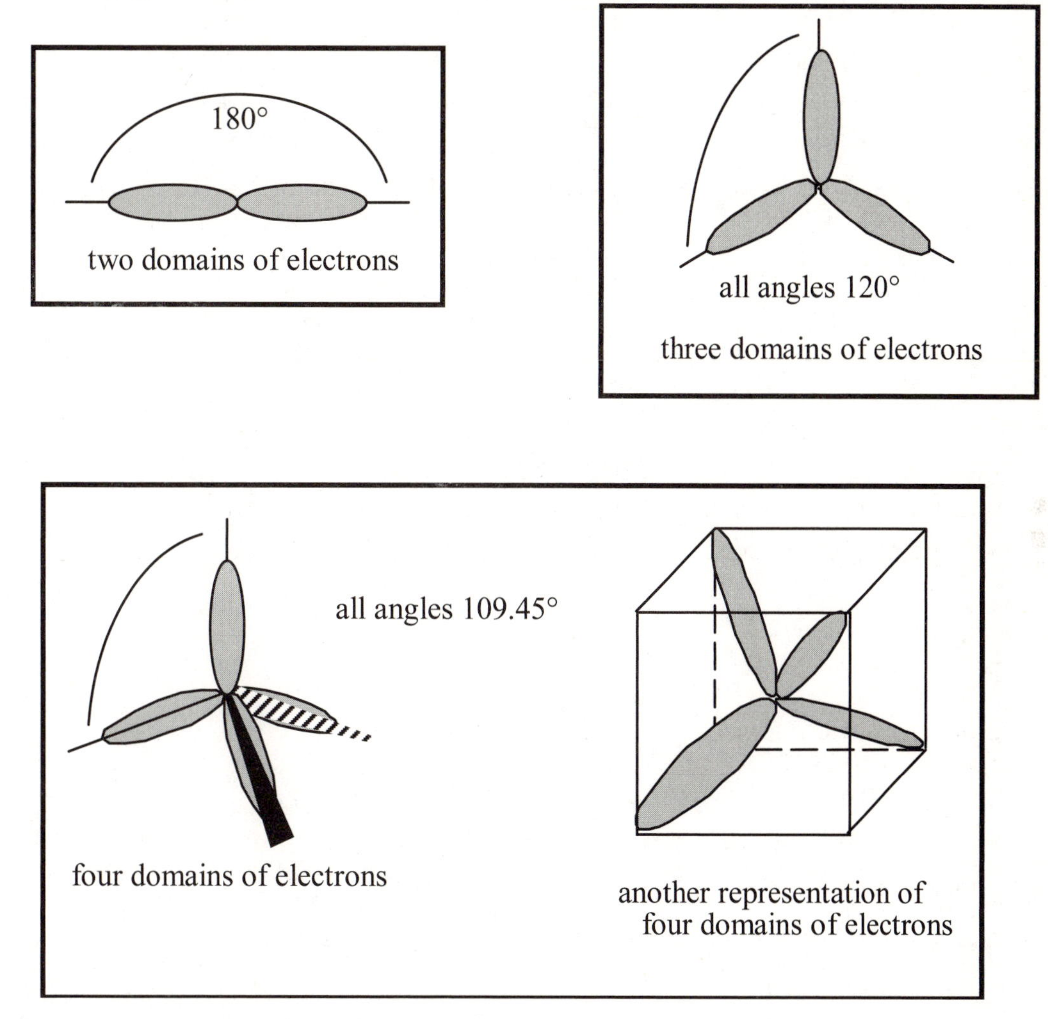

Critical Thinking Questions:

7. Based on Figure 1, what bond angle is expected for a molecule containing:
   a. two domains of electrons?
   b. three domains of electrons?
   c. four domains of electrons?

8. Write the Lewis structure for CH_4 (you may copy it from Table 1).
    a. How many domains of electrons does the carbon atom have in CH_4?
    b. Which electron domain geometry in Figure 1 applies to CH_4?
    c. Are the calculated bond angles in agreement (see Table 1)?

9. Draw the Lewis structure for NH_3.
    a. How many electron domains does the nitrogen atom have in NH_3?
    b. Which electron domain geometry in Figure 1 applies to NH_3?
    c. Are the calculated bond angles in agreement (see Table 1)?

10. Draw the Lewis structure for H_2O.
    a. How many electron domains does the oxygen atom have in H_2O?
    b. Which electron domain geometry in Figure 1 applies to H_2O?
    c. Are the calculated bond angles in agreement (see Table 1)?

11. Draw the Lewis structure for NO_3^-.
    a. How many electron domains does the nitrogen atom have in NO_3^-?
    b. Which electron domain geometry in Figure 1 applies to NO_3^-?
    c. Are the calculated bond angles in agreement (see Table 1)?

12. Draw the Lewis structure for CO_2.
    a. How many electron domains does the carbon atom have in CO_2?
    b. Which electron domain geometry in Figure 1 applies to CO_2?
    c. Are the calculated bond angles in agreement (see Table 1)?

Information:

The names for molecular shapes are based on the position of the atoms in the molecule—not on the position of the electron domains.

Figure 2. The Lewis structure, electron domains, and molecular shape of H_2O.

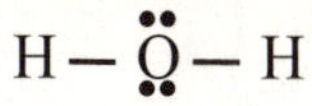

H— Ö — H Lewis structure

all angles 109.45°

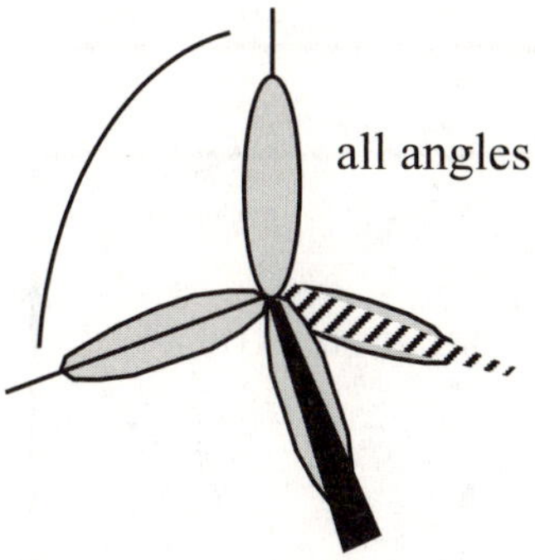

four electron domains

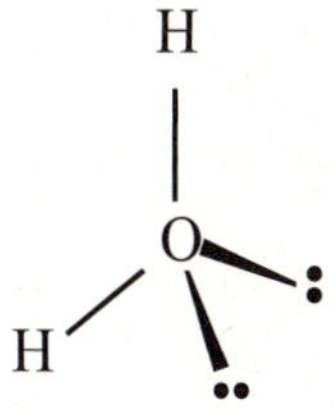

The water moluecule is said to be "bent" because the three atoms are not in a straight line. The actual bond angle, determined by experiment, is 104.5°.

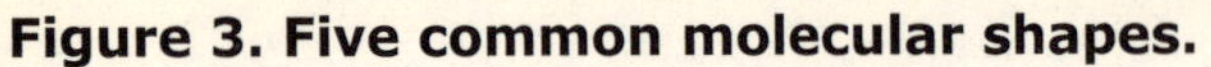
Figure 3. Five common molecular shapes.

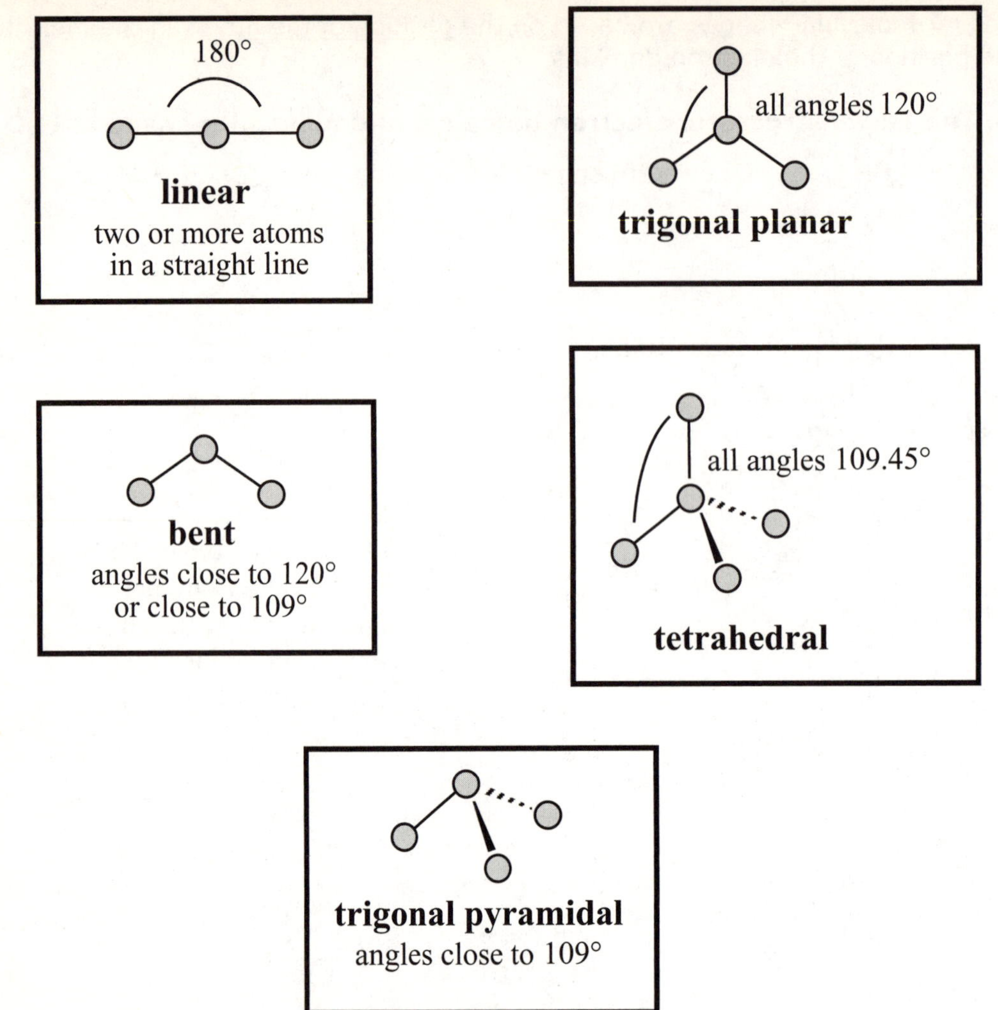

Critical Thinking Questions:

13. Considering the geometries defined in Figure 1, explain why the bond angle in bent molecules is expected to be close to 109° or close to 120°.

14. Using grammatically correct English sentences, explain how the shape of a molecule can be predicted from its Lewis structure.

Information: A Step-by-Step Method for Writing Lewis Structures

1. Write molecular skeleton (the first element in the formula, other than H, goes in the center).
2. Calculate the total number of valence electrons for all atoms in the compound (add extra for negative charges, subtract for positive charges).
3. Use electron pairs to form a bond between each pair of atoms.
4. Arrange remaining electron pairs on outer atoms to satisfy octet rule (duet rule for H/He).
5. Any electron pairs "left over" (if any) are placed on central atom.
6. If central atom does not have octet rule satisfied, share electron pairs to make multiple bonds.

Example: Draw the Lewis Structure of SO_2

Step 1. O S O

Step 2. valence electrons: 1 sulfur (6) + 2 oxygens (12) = **18** valence e^- total

Step 3. O—S—O (14 e^- remain from original 18)

Step 4. :Ö—S—Ö: (2 e^- remain)

Step 5. :Ö—S̈—Ö: (S does not have octet, so share from either O)

Step 6. :Ö═S̈—Ö:

Exercises:

1. Draw the Lewis structure for NH_3. How many electron domains does the nitrogen atom have in NH_3? _____ Make a sketch of the electron domains in NH_3. Examine the drawing for the molecular shape of H_2O given in Figure 2; make a similar drawing for NH_3. Name the shape of the NH_3 molecule, and give the approximate bond angles.

2. Draw the Lewis structure for CH_4. How many electron domains does the carbon atom have in CH_4? _____ Make a sketch of the electron domains in CH_4. Examine the drawing for the molecular shape of H_2O given in Figure 2; make a similar drawing for CH_4. Name the shape of the CH_4 molecule, and give the approximate bond angles.

3. Draw the Lewis structure for SO_2. How many electron domains does the sulfur atom have in SO_2? _____ Make a sketch of the electron domains in SO_2. Examine the drawing for the molecular shape of H_2O given in Figure 2; note that the electron domain geometry is different for SO_2, and make a similar drawing for SO_2. Name the shape of the SO_2 molecule, and give the approximate bond angles.

4. Draw the Lewis structure, sketch the molecules, predict the molecular shape, and give the bond angles for: PH_3, CO_2, SO_3^{2-}, H_3O^+, NH_2F, H_2CO. The central atom is the first atom listed (other than hydrogen).

5. Predict the bond angles around each atom designated with an arrow in the amino acid glycine, shown below.

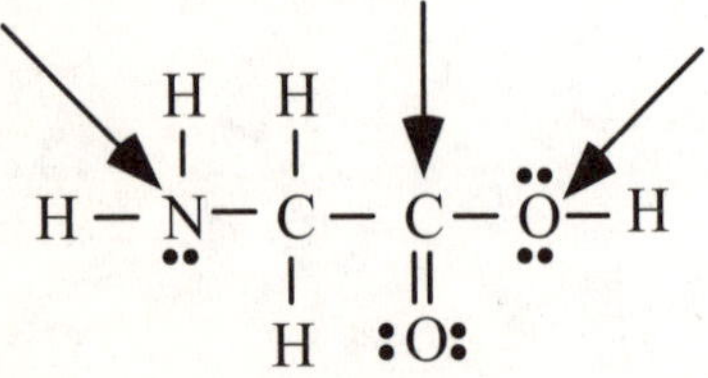

6. Read the assigned pages in your textbook and work the assigned problems.

Polar and Nonpolar Covalent Bonds

(Are atoms in molecules completely neutral?)

Model 1: Relative electronegativity values for selected elements

Element	Electronegativity
Li	1.0
H	2.2
C	2.4
O	3.5
F	4.0

The electronegativity of an atom is its ability to attract electrons in a covalent bond closer to itself. Fluorine is the most electronegative element.

Critical Thinking Questions:

1. When carbon and oxygen are covalently bonded, are the electrons in the bond attracted closer to **carbon** or **oxygen** (circle one)?
2. Which atom is at the *negatively* polarized end of a bond between carbon and oxygen?____ Which atom is at the positively polarized end? ____ Explain.
3. Which end of a bond between carbon and *lithium* is negatively polarized? ____ Explain.

Model 2: Types of bonds based on difference in electronegativity

Bond between	Chemical Formula	Electronegativity difference	Type of bond
H and H	H_2	0	nonpolar, covalent
F and F	F_2	0	nonpolar, covalent
C and H	CH_4	0.2	*slightly* polar, covalent
C and O	CO_2	1.1	polar, covalent
H and F	HF	1.8	polar, covalent
Li and F	LiF	3.0	ionic

Critical Thinking Questions:

4. Model 2 states that a H–H bond is **nonpolar**. Devise a definition for the term *nonpolar*.

5. Considering Model 2, which would you expect to be more polar—a C–O bond, or an H–F bond? Explain.

For the bond you chose, write the word "*very*" in front of "polar, covalent" in the "type of bond" column in Model 2.

6. Which would be *more* polar—a C-O bond, or a C-Li bond? Explain.

7. The Greek letter delta, δ, is often used to mean "slightly" or "partially." So, δ- would mean "partially negative," and δ+ would mean "partially positive." Place the symbols δ- or δ+ near each element below that would have a partial charge.

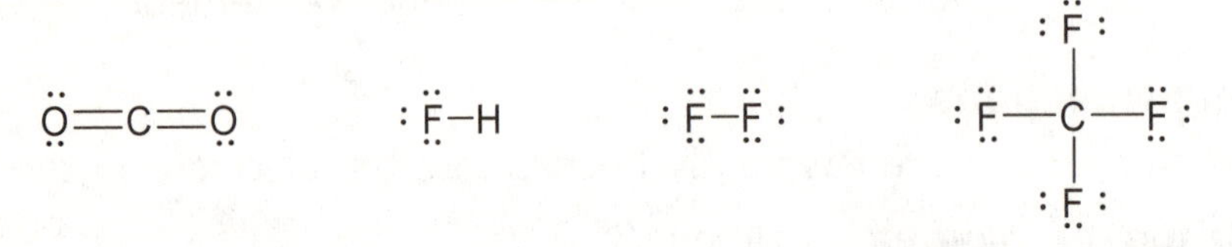

8. Based on the difference in their electronegativities (Model 2), a bond between lithium and fluorine would be extremely polarized. Which end of the bond would be *negatively* polarized— **Li** or **F** (circle one)? Explain how this is consistent with the guideline learned earlier in the course that the bond between a metal and a nonmetal is ionic.

9. We normally write the formula for lithium fluoride as a lithium ion (Li^{+}) and a fluoride ion (F^{-}). Why is this more correct that writing them as "$Li^{\delta+}$" and "$F^{\delta-}$"?

10. Explain the following statement: The ionic character of a bond increases as the electronegativity difference between the two bonded atoms increases.

11. Why might we *not* say that the bond between carbon and lithium is ionic?

12. A rule of thumb says that the closer an element is to fluorine on the periodic table, the more electronegative it is. Based on this rule, place the symbols δ- and δ+ near each element below that would have a partial charge.

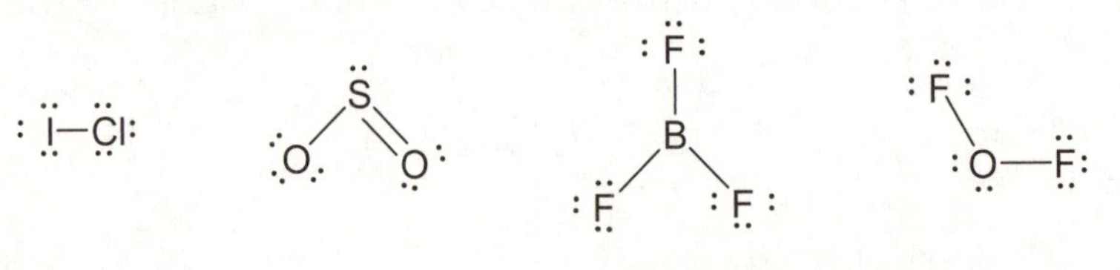

Model 3: Carbon dioxide is overall nonpolar

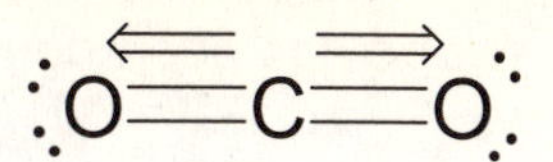

Consider the model of carbon dioxide above. Both C–O bonds are polar, with the negative ends on each oxygen. (The arrows indicate that each electronegative oxygen is pulling electrons equally toward itself.) Since both bonds are equally polarized in opposite directions, no particular side or end of the CO_2 molecule is negatively or positively polarized. In other words, the individual bond polarities "cancel out."

This type of canceling can occur whenever **all** the electrons around a central atom are in **identical** covalent bonds (to the same elements) and therefore are equally polarized.

As an analogy, consider the following: Picture the bonds as ropes, with the two oxygens pulling on the ropes in a "tug-of-war" game. Since they are "pulling" the electrons equally in opposite directions, there is no net movement of the electrons, and the molecule is **overall nonpolar**.

Critical Thinking Questions:

13. Place the symbols δ- or δ+ near each element in Model 3 that would have a partial charge.

14. Three molecules in CTQ 7 have polar *bonds*, but only one *molecule* is polar. Explain.

15. One molecule in CTQ 12 is *nonpolar*. Circle it. Explain why it is overall nonpolar.

Exercises:

1. Complete the table below. The first one has been done as an example.

Formula	Total Valence Electrons	Lewis Structure	Shape Around Most Central Atom	Approximate Bond Angles	Ion, Polar Molecule, or Nonpolar Molecule
CF_4	32	:F̤̈: / | / :F̤̈—C—F̤̈: / | / :F̤̈:	tetrahedral	109.5°	nonpolar molecule
SO_4^{-2}					
CH_2O					
$CHCl_3$					
CH_3OH (bond the last H to the O)					
HCN					
NO_2^-					

2. Go back to your Lewis structures in Exercise 1 and indicate the bond polarities by placing δ− or δ+ near all the atoms. Exception: The polarity of a C-H bond is so slight that it is normally ignored.
3. Read the assigned pages in your textbook and work the assigned problems.

The Mole Concept[*]

(What is a mole?)

Model 1: Some selected conversion factors and information

1 dozen items = 12 items

1 score of items = 20 items

1 myriad items = 10,000 items

1 mole of items= 6.022×10^{23} items (Avogadro's Number)

One elephant has one trunk and four legs.

One methane molecule, CH_4, contains one carbon atom and four hydrogen atoms.

Critical Thinking Questions:

Use scientific notation if any answer is a very large or very small number. Include units with all answers.

1. How many trunks are found in one dozen elephants?

2. How many legs are found in one dozen elephants?

3. How many trunks are found in one score of elephants?

4. How many legs are found in one myriad elephants?

5. How many trunks are found in one mole of elephants?

6. How many carbon atoms are found in one dozen methane molecules?

7. How many hydrogen atoms are found in one myriad methane molecules?

---

[*] Adapted from ChemActivity 28, Moog, R.S. and Farrell, J.J. *Chemistry: A Guided Inquiry*, 3rd ed., Wiley, 2006, pp. 158-160.

8. How many elephants are there in one mole of elephants?

9. How many trunks are found in one-half mole of elephants?

10. How many legs are found in one mole of elephants?

11. How many carbon atoms are found in one mole of methane molecules?

12. How many hydrogen atoms are found in one-half mole of methane molecules?

13. How many methane molecules are there in one mole of methane?

Model 2: The relationship between average atomic mass and moles

The mass of 1 mole of any pure substance is equal to the average atomic mass expressed in grams of that substance.

For example: The average atomic mass of a helium atom is 4.003 amu (atomic mass units).

Therefore: 4.003 g He = 1 mol He

Critical Thinking Questions:

14. What is the average mass (in amu) of one carbon atom?

15. What is the mass (in grams) of one mole of carbon atoms?

16. What is the average mass (in amu) of one methane molecule?

17. What is the mass (in grams) of one mole of methane molecules?

18. Use a grammatically correct English sentence to describe how the mass in amu of one molecule of a compound is related to the mass in grams of one mole of that compound.

Exercises:

Unless otherwise stated, calculate all mass values in grams.

1. Consider CTQ 1. A unit plan for this problem would be dozens → elephants → trunks. Since you know there are 12 elephants in 1 dozen elephants and 1 trunk per elephant, you could set up the problem using conversion factors as follows:

$$1 \text{ dozen elephants} \times \frac{12 \text{ elephants}}{1 \text{ dozen elephants}} \times \frac{1 \text{ trunk}}{1 \text{ elephant}} = 12 \text{ trunks}$$

   Complete similar unit conversions, showing all work, for CTQs 2-13.

2. If you measure out 69.236 g of lead, how many atoms of lead do you have? Make a unit plan first. Show work.

3. Consider 1.00 mole of dihydrogen gas, H_2. How many dihydrogen molecules are present? How many hydrogen atoms are present? What is the mass of this sample?

4. Ethanol has a molecular formula of CH_3CH_2OH.

   a. What is the average mass of one molecule of ethanol?

   b. What is the mass of 1.000 moles of ethanol?

   c. What is the mass of 0.5623 moles of ethanol, CH_3CH_2OH?

   d. How many moles of ethanol are present in a 100.0 g sample of ethanol?

   e. How many moles of each element (C, H, O) are present in a 100.0 g sample of ethanol?

   f. How many grams of each element (C, H, O) are present in a 100.0 g sample of ethanol?

5. How many moles of carbon dioxide, CO_2, are present in a sample of carbon dioxide with a mass of 254 grams? How many moles of O atoms are present?

6. Indicate whether each of the following statements is true or false, and explain your reasoning.

   a. One mole of NH_3 weighs more than one mole of H_2O.

   b. There are more carbon atoms in 48 grams of CO_2 than in 12 grams of diamond (a form of pure carbon).

   c. There are equal numbers of nitrogen atoms in one mole of NH_3 and one mole of N_2.

   d. The number of Cu atoms in 100 grams of Cu(s) is the same as the number of Cu atoms in 100 grams of copper(II) oxide, CuO.

   e. The number of Ni atoms in 100 moles of Ni(s) is the same as the number of Ni atoms in 100 moles of nickel(II) chloride, $NiCl_2$.

   f. There are more hydrogen atoms in 2 moles of NH_3 than in 2 moles of CH_4.

Balancing Chemical Equations

(What stays the same and what may change in a chemical reaction?)

Model: Atoms are conserved in chemical reactions

Atoms are neither created nor destroyed when chemical reactions occur. Two balanced chemical reactions (or equations) are given below. In a chemical reaction equation, reactant atoms are shown on the left side of the arrow and product atoms are shown on the right.

$$2\ H_2\ (g) + O_2\ (g) \rightarrow 2\ H_2O\ (g) \qquad (1)$$

$$Fe_2O_3\ (s) + 2\ Al\ (s) \rightarrow 2\ Fe\ (l) + Al_2O_3\ (s) \qquad (2)$$

Critical Thinking Questions:

1. Considering reaction (1) above:
   a. What are the reactants?

   b. What are the products?

2. What does the arrow represent in a chemical reaction?

3. How many H atoms are represented:
   a. On the reactant side of reaction (1)?
   b. On the product side?
4. Compare the number of atoms of each element (H, O) on the reactant and product sides of equation (1). Are these numbers **the same** or **different** (circle one)?
5. Compare the number of atoms of each element (Fe, Al, O) on the reactant and product sides of equation (2). Are these numbers **the same** or **different** (circle one)?
6. For each H_2 molecule consumed in reaction (1), how many H_2O molecules are produced?

7. For each O_2 molecule consumed in reaction (1), how many H_2O molecules are produced?

8. For each mole of oxygen molecules consumed in reaction (1), how many moles of water molecules are produced?

9. Considering your answers to CTQs 7 and 8, what two things can be indicated by the coefficient (the number) in front of the chemical formulas?

10. For each mole of H_2O molecules that are produced in reaction (1), how many moles of O_2 molecules are required?

11. Describe in a complete sentence or two how you arrived at your answer to CTQ 10.

12. One of the interpretations for the coefficients that you listed in CTQ 9 is consistent with your answer to CTQ 10, and one is not. Explain.

13. Considering reaction (1), what it the total number of molecules of water produced from two molecules of hydrogen and one molecule of oxygen?_____

14. Is the total number of **molecules** identical on the reactant and product sides of these balanced equations? _____

15. Explain how your answer to CTQ 14 can be consistent with the idea that atoms are neither created nor destroyed when chemical reactions take place.

Exercises:

1. Balance the equations by adding **coefficients** in front of each reactant and product.

   a. $Cr\ (s) + \quad S_8\ (s) \rightarrow \quad Cr_2S_3\ (s)$

   b. $Fe_3O_4\ (l) + \quad CO\ (g) \rightarrow \quad FeO\ (l) + \quad CO_2\ (g)$

   c. $CH_4\ (g) + \quad O_2\ (g) \rightarrow \quad CO_2\ (g) + \quad H_2O\ (g)$

2. For each mole of S_8 molecules consumed in the reaction in Exercise 1, how many moles of Cr_2S_3 molecules are produced?

3. In the balanced reaction: $2\ CO\ (g) + O_2\ (g) \rightarrow 2\ CO_2\ (g)$

   a. How many moles of CO_2 can be produced from 4 moles CO and 2 moles O_2?

   b. How many grams of CO_2 would this be?

4. Add molecular (spacefilling) pictures to the boxes below in the correct ratio so that it represents reaction (1) in the Model. For example, could be used to represent one molecule of H_2.

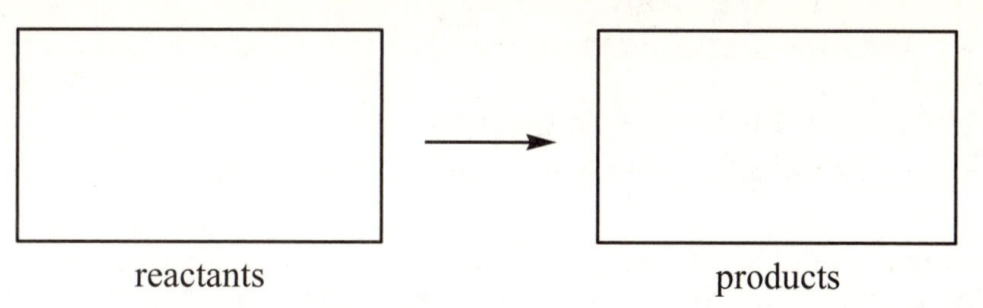

5. Use molecular (spacefilling) pictures in the correct ratio to represent the reaction in Exercise 1a. Be sure that the representations for Cr atoms and S atoms are distinguishable. (You may wish to include a key.)

6. Read the assigned pages in your text, and work the assigned problems.

Predicting Binary Reactions

(When will precipitation or neutralization reactions occur?)

Table 1: Solubility rules for ionic compounds

1. A compound containing a cation from Group 1 (alkali metal) or the ammonium ion (NH_4^+) is likely soluble.
2. A compound containing an anion from group 17 (halides) is likely soluble.
   - **Exceptions:** Ag^+, Hg_2^{2+}, and Pb^{2+} halides are insoluble.
3. A compound containing nitrate (NO_3^-), acetate ($CH_3CO_2^-$), or sulfate (SO_4^{2-}) is likely soluble.
   - **Exceptions:** The sulfates of Ba^{2+}, Hg_2^{2+}, and Pb^{2+} are insoluble.
4. **Any other** ionic compounds are likely insoluble.

Unless your instructor tells you otherwise, you are not required to memorize the information in Table 1. You may refer to the table when needed.

Critical Thinking Questions:

1. Write the chemical formula for each compound.

   a. barium nitrate ______________________

   b. sodium sulfate ______________________

   c. lead(II) nitrate ______________________

   d. sodium iodide ______________________

   e. sodium hydroxide ______________________

   f. sulfuric acid ______________________

   g. ammonium chloride ______________________

   h. lead(II) iodide ______________________

2. Suppose that each compound in CTQ 1 is placed in water. Using Model 1 for guidance, place the symbol (aq) or (s) after each formula to indicate its solubility in water.

Information: Binary reactions (Double replacement reactions)

Chemical reactions occur between two reactants whenever two **electrolytes** can combine to form a **nonelectrolyte** (*i. e.,* an insoluble solid or a covalent molecule such as water). Consider the reaction of barium hydroxide and sulfuric acid, shown below:

$$Ba(OH)_2\,(aq) + H_2SO_4\,(aq) \rightarrow BaSO_4\,(s) + 2\,H_2O\,(l)$$

The reactants are strong electrolytes (strong base and strong acid), but the products are both nonelectrolytes. This reaction is both a *neutralization* and a *precipitation*.

⇒ Neutralization: acid + base → salt + water
⇒ Precipitation: formation of an insoluble solid

If one or more nonelectrolytes can be formed, a reaction will occur.

If no nonelectrolytes can be formed, no reaction occurs.

Critical Thinking Questions:

Write a balanced molecular equation starting with each set of reactants below. Include the phase labels. Write NR for no reaction.

Example: Aqueous **barium nitrate** and aqueous **sodium sulfate**.

- Switch the "partners" to make **barium sulfate** and **sodium nitrate**.
- **Balance the ionic charges** to get the right formula for each product (as in CTQ 1).

$$Ba(NO_3)_2 + Na_2SO_4 \rightarrow BaSO_4 + NaNO_3$$

- Use the solubility rules in Table 1 to **assign the phase** for each product. (If all products are electrolytes, no reaction occurs, and you may stop here; write *NR*).

$$Ba(NO_3)_2\ (aq) + Na_2SO_4\ (aq) \rightarrow BaSO_4\ (s) + NaNO_3\ (aq)$$

- Finally, **balance the equation** by assigning coefficients.

$$Ba(NO_3)_2\ (aq) + Na_2SO_4\ (aq) \rightarrow BaSO_4\ (s) + \mathbf{2}\ NaNO_3\ (aq)$$

3. Aqueous lead(II) nitrate and aqueous sodium iodide.

4. Aqueous sodium hydroxide and aqueous sulfuric acid.

5. Aqueous ammonium chloride and aqueous potassium nitrate.

Exercises:

1. Using the solubility rules, write a balanced molecular equation starting with each set of reactants below. Include the phase labels.
    a. Aqueous silver nitrate and aqueous sodium phosphate (silver always has a +1 charge in ionic compounds).

    b. Aqueous calcium nitrate and aqueous potassium carbonate.

    c. Aqueous copper(II) nitrate and aqueous potassium hydroxide.

    d. Aqueous zinc chloride and aqueous sodium sulfide (zinc always has a +2 charge in ionic compounds).

    a. Aqueous lead(II) nitrate and aqueous sodium phosphate.

    b. Aqueous magnesium sulfate and aqueous sodium hydroxide.

2. Read the assigned pages in your text, and work the assigned problems.

Oxidation-Reduction Reactions

(What happens when electrons are transferred between atoms?)

Information:

Oxidation-reduction reactions (often called "redox" reactions) are those in which *electrons are transferred between atoms*. Oxidation and reduction always occur together, since an atom that loses electrons must give them to an atom that gains the electrons.

A way to keep track of electrons in chemical reactions is by assigning **oxidation numbers** to each element in a reaction. The oxidation number is the charge an atom in a substance would have if the electron pairs in each covalent bond belonged to the more electronegative atom.

An atom that...	is said to be...	and its oxidation number...
gains electrons	reduced	decreases
loses electrons	oxidized	increases

You are responsible for recognizing (but not predicting) redox reactions.

Table 1. General rules for assigning oxidation numbers

1. Pure **elements** (not in compounds) have an oxidation number of 0. This includes diatomic elements like H_2 or O_2.
2. The oxidation number of an **ion** equals its charge. For example:
   - group 1 (alkali metals) → +1
   - group 2 (alkaline earth) → +2
3. In **molecules, F** is –1; **Cl**, **Br**, **I** are –1 if bonded to less electronegative atom
4. **Oxygen** is nearly always –2
5. **Hydrogen** nearly always +1
6. Figure any others by calculation. The sum of oxidation numbers of the atoms of a molecule equals zero; the sum for an ion should equal its overall charge.
   - For example, in CO_2, the oxidation number of each O is –2, so the oxidation number of C must be +4.

Critical Thinking Questions:

1. Assign an oxidation number to each atom individually in the following substances.

   a. CCl_4 b. CH_4 c. HNO_3 d. $Ca(NO_3)_2$ e. $K_2Cr_2O_7$ f. SO_4^{2-}

2. Which element is oxidized and which is reduced in the following reaction?

   $$5\ Fe^{2+}(aq) + MnO_4^-(aq) + 8\ H^+(aq) \rightarrow Mn^{2+}(aq) + 5\ Fe^{3+}(aq) + 4\ H_2O(l)$$

3. Consider the balanced chemical equation:

   $$5\ H_2C_2O_4(aq) + 2\ MnO_4^-(aq) + 6\ H^+(aq) \rightarrow 10\ CO_2(g) + 2\ Mn^{2+}(aq) + 8\ H_2O(l)$$

   When one mole of oxalic acid, $H_2C_2O_4$, reacts, _____ moles of permanganate, MnO_4^-, will be consumed.

   a) 1 b) 2 c) 2/5 d) 5/2

4. Is the reaction in CTQ 3 a redox reaction? Explain your answer in terms of oxidation numbers.

Information: Redox reactions in biological systems (organic molecules)

It is often difficult to assign oxidation numbers in complex molecules, but there is a shortcut. In these cases, a molecule that undergoes **addition of oxygen atoms** or **loss of hydrogen atoms** is **oxidized**. If oxygen atoms are **removed** (*i. e.,* a product) or hydrogen atoms are **added** (*i. e.,* a reactant), the molecule is **reduced**. However, if H_2O (water) is added or removed, this is **not** a redox reaction. Often, we do not balance these reactions, so *the added or lost hydrogen or oxygen may not be shown*.

Example: $CH_3OH \rightarrow CH_2O$ (+ H_2) CH_3OH is oxidized (loss of 2 H)

Example: C_4H_{10} (+ H_2) $\rightarrow C_4H_{12}$ C_4H_{10} is reduced (gain of 2 H)

Example: $C_4H_{10} \rightarrow C_4H_{12}O$ neither (addition of H_2O, not balanced)

Example: $CH_2O \rightarrow CH_2O_2$ oxidation (addition of O, not balanced)

Critical Thinking Questions:

5. Glucose ($C_6H_{12}O_6$) is used for energy in cells in cellular respiration by the net reaction shown:

$$C_6H_{12}O_6 + 6\ O_2 \rightarrow 6\ CO_2 + 6\ H_2O$$

Is glucose oxidized, reduced, or neither? Explain both in terms of the oxidation number of carbon and the shortcut for organic reactions.

6. When linoleic acid, an unsaturated fatty acid, reacts with hydrogen, it becomes a saturated fatty acid.

$C_{18}H_{32}O_2 \rightarrow C_{18}H_{36}O_2$ (not balanced)

This process, called hydrogenation, is used to make shortening ("Crisco") out of vegetable oil. In this hydrogenation, is linoleic acid oxidized or reduced? Explain.

Exercises:

1. Assign oxidation numbers to each atom in the reactions below. Then tell which atom is oxidized and which is reduced in each reaction.

   a. $CuO\ (s) + H_2\ (g) \rightarrow Cu\ (s) + H_2O\ (g)$

   b. $2\ CO\ (g) + O_2\ (g) \rightarrow 2\ CO_2\ (g)$

2. Is the reaction in Exercise 1a an **oxidation** or a **reduction** **(circle one)** of copper(I) oxide? Explain how you can tell **without** assigning oxidation numbers.

3. Is the reaction in Exercise 1b an **oxidation** or **reduction** of carbon monoxide? Explain how you can tell **without** assigning oxidation numbers.

4. When ethanol is ingested and metabolized in the liver, it is first converted to acetaldehyde, and then to acetic acid, as shown in the Lewis structures below.

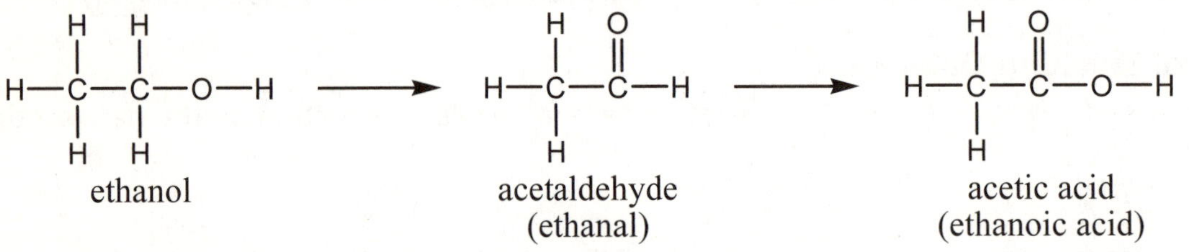

   a. Is the conversion of ethanol to acetaldehyde an **oxidation** or **reduction** **(circle one)**? Explain how you can tell.

   b. Is the conversion of acetaldehyde to acetic acid an **oxidation** or **reduction** **(circle one)**? Explain how you can tell.

5. According to the balanced chemical equation:

   $5\ H_2C_2O_4(aq) + 2\ KMnO_4(aq) + 6\ H^+(aq) \rightarrow 10\ CO_2(g) + 2\ Mn^{2+}(aq) + 8\ H_2O(l) + 2\ K^+(aq)$

   How many **moles** of potassium permanganate, $KMnO_4$, will exactly react with 0.3500 **grams** of oxalic acid, $H_2C_2O_4$, in acidic solution. Show work. (Note that this is the same reaction as in CTQ 3, but with the spectator ion potassium included.) How many **grams** of $KMnO_4$ would this be?

6. Read the assigned pages in your text, and work the assigned problems.

*Mass Relationships (Stoichiometry)**

(How much can you make?)

Model 1: The S'more:

A delicious treat known as a S'more is constructed with the following ingredients and amounts:

1 graham cracker
1 chocolate bar
2 marshmallows

At a particular store, these items can be obtained only in full boxes, each of which contains one gross of items. A gross is a specific number of items, analogous (but not equal) to one dozen. The boxes of items have the following net weights (the weight of the material inside the box):

box of graham crackers	9.0 pounds
box of chocolate bars	36.0 pounds
box of marshmallows	3.0 pounds

Critical Thinking Questions:

1. If you have a collection of 100 graham crackers, how many chocolate bars and how many marshmallows do you need to make S'mores using all the graham crackers?

2. If you have a collection of 1000 graham crackers, 800 chocolate bars, and 1000 marshmallows:

   a. How many S'mores can you make?

   b. What (if anything) will be left over, and how many of that item will there be?

Information:

Chemists refer to the reactant which limits the amount of product which can be made from a given collection of original reactants as the **limiting reagent** or **limiting reactant.**

---

* Adapted from ChemActivity 30, Moog, R.S. and Farrell, J.J. *Chemistry: A Guided Inquiry.* Wiley, 3rd ed., 2006, pp. 168-171.

Critical Thinking Questions:

3. Identify the limiting reagent in CTQ 2.

4. Based on the information given:
   a. Which of the three ingredients (a graham cracker, a chocolate bar, or a marshmallow) weighs the most?
   b. Which weighs the least?

   Explain your reasoning.

5. If you have 36.0 pounds of graham crackers, 36.0 pounds of chocolate bars, and 36.0 pounds of marshmallows:
   a. Which item do you have the most of?
   b. Which item do you have the least of?

   Explain your reasoning.

6. If you attempt to make S'mores from the material described in CTQ 5:
   a. What will the limiting reagent be?
   b. How many gross of S'mores will you have made?
   c. How many gross of each of the two leftover items will you have?
   d. How many pounds of each of the leftover items will you have?
   e. How many pounds of S'mores will you have?

7. Using G as the symbol for graham cracker, Ch for chocolate bar, and M for marshmallow, write a "balanced chemical equation" for the production of S'mores.

8. Explain in complete sentences why it is not correct to state that if we start with 36 pounds each of G, Ch, and M then we should end up with 3 x 36 = 108 pounds of S'mores.

Model 2: Water

Water can be made by burning hydrogen in oxygen, as shown below:

$$2\ H_2 + O_2 \rightarrow 2\ H_2O$$

A mole is a specific number of items, analogous (but not equal) to one dozen. The reactants in the equation have the following molar masses (the weight of the material in a mole of material):

mole of H_2 molecules	2.0 grams
mole of O_2 molecules	32.0 grams

Critical Thinking Questions:

9. Based on the information given:

   a. Which of the two ingredients (hydrogen molecules or oxygen molecules) weighs the most?

   b. Which weighs the least?

   Explain your reasoning.

10. If you have a collection of 100 hydrogen molecules, how many oxygen molecules do you need to make water with all of the hydrogen molecule?

11. If you have 32.0 grams of hydrogen molecules and 32.0 grams of oxygen molecules,

    a. which item do you have the most of?

    b. which item do you have the least of?

    Explain your reasoning.

12. If you attempt to make water from the material described in CTQ 11:

   a. What is the limiting reagent?

   b. How many moles of water molecules will you have made?

   c. How much does one mole of water molecules weigh? Explain how you can determine this.

   d. Based on your answer to (b), how many grams of water will you have?

   e. A student performed the reaction in CTQ 11 and obtained only 34.0 grams of water. What percentage of the total possible amount of water did he obtain?

Information: What if the answers are not in whole numbers? (Example)

Problem: Methane gas (CH_4) burns in oxygen (O_2) to produce carbon dioxide and water. How many grams of CO_2 can be produced from 2.0 g of methane and 2.0 g of oxygen?

- First, write the two products and **balance** the equation:

  ___ CH_4 + ___ O_2 $\rightarrow$

- Then, using the **molar mass** (g/mol) for each reactant, see how many moles of each reactant you have:

$$2.0\text{ g CH}_4 \cdot \frac{1\text{ mol CH}_4}{16.04\text{ g CH}_4} = 0.1247\text{ mol CH}_4$$

$$2.0\text{ g O}_2 \cdot \frac{1\text{ mol O}_2}{32.00\text{ g O}_2} = 0.0625\text{ mol O}_2$$

We need to determine which reagent will be completely used up. Look at the balanced equation again. It shows us that we need two times as many O_2 moles as CH_4 moles. Do we have this (is 0.0625 mol twice as much as 0.1247 mol)? **Yes** or **no** (circle one). So the **CH_4** or **O_2** (circle one) will be completely used up (the **limiting reagent**). The calculation of the moles of the reagent that is **not** limiting is now useless. Draw a large "X" through that calculation above to indicate that you will not use it.

Now, the question asks how much CO_2 is produced. Look at the balanced equation yet again. It shows that the moles of CO_2 produced will be half the number of moles of O_2 used up. Since O_2 was the limiting reagent, you can multiply the moles of O_2 by ½, or use a conversion factor, as follows:

$$0.0625\text{ mol O}_2 \cdot \frac{1\text{ mol CO}_2}{2\text{ mol O}_2} = 0.03125\text{ mol CO}_2$$

See how the coefficients from the equation make a conversion factor ("½") for you? (Include the units!) Now just use the **molar mass** of CO_2 to change the moles into grams, and you're done!

$$0.03125 \text{ mol } CO_2 \cdot \frac{44.01 \text{ g } CO_2}{1 \text{ mol } CO_2} = 1.4 \text{ g } CO_2$$

Part 2: If 1.0 g of CO_2 are produced in the above reaction, what is the percent yield?

The 1.4 g is how much you can **theoretically** get. But we only got 1.0 g, so maybe some gas escaped, or perhaps our reaction was incomplete. Our **percent yield** is:

$$\frac{1.0 \text{ g produced}}{1.4 \text{ g theoretical}} \times 100 = 71\% \text{ yield}$$

Exercises:

1. The thermite reaction, once used for welding railroad rails, is often used for an exciting chemistry demonstration because it produces red-hot molten iron. The reaction is:

   Fe_2O_3 (s) + 2 Al (s) $\rightarrow$ 2 Fe (l) + Al_2O_3 (s)

   If you start with 50.0 g of iron(III) oxide and 25.0 g of aluminum, what is the limiting reagent? What is the maximum mass of aluminum oxide that could be produced? How much aluminum oxide would be produced if the yield is 93%?

2. How much aspirin can be made from 100.0 g of salicylic acid and 100.0 g of acetic anhydride? If 122 g of aspirin are obtained by the reaction, what is the percent yield?

$C_7H_6O_3$	+	$C_4H_6O_3$	$\rightarrow$	$C_9H_8O_4$	+	$HC_2H_3O_2$
salicylic acid		acetic anhydride		aspirin		acetic acid

3. Complete ChemWorksheet 1: Stoichiometry Practice Worksheet 1.
4. Read the assigned pages in your text, and work the assigned problems.

Thermochemistry

(What does heat have to do with chemistry?)

Suggested demonstration: Exothermic and endothermic reactions

Information:

The chemicals in a reaction **system** can either absorb heat energy from the **surroundings** or release heat energy to the surroundings. A reaction in which heat energy moves from the system to the surroundings (and feels warm) is said to be exothermic. A reaction in which heat energy moves from the surroundings to the system (and feels cool) is said to be endothermic.

Heat energy is measured in calories (English system) or Joules (metric system).

1 cal = 4.184 joules (exact)

Dietary calories reported on food products are actually kilocalories, written as "kcal" or "Cal" (with a capital "C").

Model 1: Heating curve of a pure substance

The graph (heating curve) below shows the temperature of 19 g of a pure substance that is heated by a constant source supplying 500.0 calories per minute.

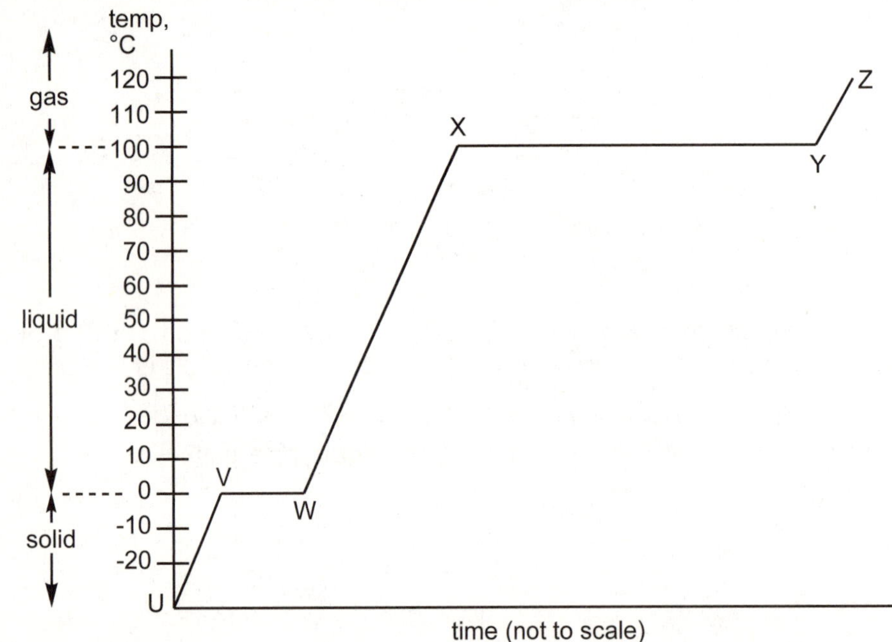

Regions: UV = 0.76 min, VW = 3.04 min, WX = 3.8 min, XY = 20.5 min, YZ = 0.63 min

Critical Thinking Questions:

1. At the coldest temperature (point U), what is the physical state of the pure substance?

   Circle one: **solid** **liquid** **gas**.

2. Identify the region on the graph (UV, VW, *etc.*) that corresponds to the pure substance:
   a. being warmed while remaining a solid _____
   b. being warmed while remaining a liquid _____
   c. being warmed while remaining a gas _____
   d. changing from a solid to a liquid _____
   e. changing from a liquid to a gas _____
3. What is the boiling point of the liquid? _____
4. What is the melting point of the liquid? _____
5. Why is the temperature not changing from point V to W, even though heat is being added?

6. Why is the temperature not changing from point X to point Y?

7. How many calories were needed to melt the solid?

8. From your answer to CTQ 7, calculate the calories used per gram (cal/g) of this substance in order to melt it. Include units. This quantity is known as the **heat of fusion** of the substance.

9. How many calories were needed to change the liquid to a gas?

10. From your answer to CTQ 9, calculate the calories used per gram (cal/g) of this substance to vaporize it. Include units. This quantity is known as the **heat of vaporization** of the substance.

11. How many calories were needed to warm the liquid from point W to point X?

12. How many degrees Celsius did the temperature **change** from point W to point X?

13. From your answers to CTQs 11 and 12, calculate the calories used per gram of this substance per degree Celsius ($^{cal}/_{g\,°C}$) to change its temperature. Include units. This quantity is known as the **specific heat capacity** (or "specific heat") of the substance.

14. Where on the curve do the molecules have the highest kinetic energy? ___________

15. What do you think is the identity of the substance being heated? ___________

16. Complete the definitions of the terms in **boldface** in CTQs 8, 10, and 13.

    a. The heat of fusion of a substance is

    b. The heat of vaporization of a substance is

    c. The specific heat capacity of a substance is

Model 2: Specific heat capacity (or "specific heat")

The heat energy in calories (q) needed to raise the temperature of a substance may be calculated using the following equation:

$$q \text{ (heat in calories)} = s \text{ (specific heat, } \frac{\text{cal}}{\text{g°C}}) \times (\text{mass in grams}) \times (\Delta T \text{ in °C})$$

or simply:

$$q = s \cdot m \cdot \Delta T$$

17. How much heat energy is required to completely change 25 grams of **liquid water** at 0°C to water at body temperature (37°C)?

18. Consider how much heat energy is required to completely change 25 grams of **ice** at 0°C to **water** at 0°C.

    a. Circle the constant that is needed to perform this calculation:

    heat of fusion **heat of vaporization** **specific heat**

    b. Perform the calculation.

19. How much total heat energy is required to completely change 25 grams of ice at 0°C to water AND also raise its temperature to body temperature (37°C)?

20. A hot iron skillet (178°C) weighing 1.51 kg is sitting on a stove. How much heat energy (in joules) must be removed to cool the skillet to room temperature, 21°C? The specific heat of iron is 0.450 J/(g·°C).

21. Where does the heat energy in CTQ 20 go? ______________________

22. Is the process in CTQ 20 exothermic or endothermic? ______________________

Exercises:

1. Convert your answer to CTQ 19 into kilocalories. Would eating ice be a good way to lose weight? Explain why or why not.

2. It takes 5.85 kJ of heat energy to raise the temperature of 125.6 g of mercury from 20.0°C to 28.3°C. Calculate the specific heat capacity of mercury.

3. When 23.6 g of calcium chloride were dissolved in 300 mL of water in a calorimeter, the temperature of the water rose from 25.0°C to 38.7°C. What is the heat energy change in kcal for this process? [The specific heat of H_2O = 1.00 cal/g °C.]

4. In Exercise 3 you calculated the heat energy produced by dissolving 23.6 g of calcium chloride in water. How many **moles** of calcium chloride is this? How much heat energy is produced **per mole** of calcium chloride that dissolves in water? Report your answer in **kilocalories/mole** (kcal/mol).

5. Read the assigned pages in your text, and work the assigned problems.

Equilibrium[*]

(Do reactions really ever stop?)

Model 1: The conversion of *cis*-2-butene to *trans*-2-butene.

Consider a simple chemical reaction where the forward reaction occurs in a single step and the reverse reaction occurs in a single step:

A qwe B

The following chemical reaction, where *cis*-2-butene is converted into *trans*-2-butene, is an example.

cis-2-butene   *trans*-2-butene

In this example, one end of a *cis*-2-butene molecule rotates 180° to form a *trans*-2-butene molecule. Rotation around a double bond rarely happens at room temperature because the collisions are not sufficiently energetic to weaken the double bond. At higher temperatures, around 400°C for *cis*-2-butene, collisions are sufficiently energetic and an appreciable reaction rate is detected.

Critical Thinking Questions:

1. Make a model of *cis*-2-butene with a modeling kit. (In most model kits, black = C, white = H. Use short bonds for single bonds and two of the longer, flexible bonds for double bonds.) What must be done to convert *cis*-2-butene to *trans*-2-butene?

2. Make a model of *trans*-2-butene with a modeling kit. What must be done to convert *trans*-2-butene to *cis*-2-butene?

3. A large number of *cis*-2-butene molecules is placed in a container.

   a. Predict what will happen if these molecules are allowed to stand at room temperature for a long time.

   b. Predict what will happen if these molecules are allowed to stand at 400°C for a long time.

[*] Adapted from ChemActivity 37, Moog, R.S. and Farrell, J.J. *Chemistry: A Guided Inquiry*, 3rd ed., Wiley, 2006, pp. 205-209.

Model 2. The number of molecules as a function of time.

Consider the simple reaction:

A qwe B

The system is said to be at equilibrium when the concentrations of reactants and products stop changing.

Imagine the following hypothetical system. Exactly 10,000 A molecules are placed in a container which is maintained at 800°C. We have the ability to monitor the number of A molecules and the number of B molecules in the container at all times. We collect the data at various times and compile Table 1.

Table 1. Number of A and B molecules as a function of time

Time (seconds)	Number of A Molecules	Number of B Molecules	Number of A Molecules that React in Next Second	Number of B Molecules that React in Next Second	Number of A Molecules Formed in Next Second	Number of B Molecules Formed in Next Second
0	10000	0	2500	0	0	2500
1	7500	2500	1875	250	250	1875
2	5875	4125	1469	413	413	1469
3	4819	5181	1205	518	518	1205
4	4132	5868	1033	587	587	1033
5	3686	6314	921	631	631	921
6	3396	6604	849	660	660	849
7	3207	6793	802	679	679	802
8	3085	6915	771	692	692	771
9	3005	6995	751	699	699	751
10	2953	7047	738	705	705	738
11	2920	7080	730	708	708	730
12	2898	7102	724	710	710	724
13	2884	7116	721	712	712	721
14	2874	7126	719	713	713	719
15	2868	7132	717	713	713	717
16	2864	7136	716	714	714	716
17	2862	7138	715	714	714	715
18	2860	7140	715	714	714	715
19	2859	7141	715	714	714	715
20	2858	7142	715	714	714	715
21	2858	7142	714	714	714	714
22	2858	7142	714	714	714	714
23	2857	7143	714	714	714	714
24	2857	7143	714	714	714	714
25	2857	7143	714	714	714	714
30	2857	7143	714	714	714	714
40	2857	7143	714	714	714	714
50	2857	7143	714	714	714	714

Critical Thinking Questions:

4. During the time interval 0–1 s:
   a. How many A molecules react? ______
   b. How many B molecules are formed? ______
   c. Why are these two numbers equal?

5. During the time interval 10–11 s:
   a. How many B molecules react? _____
   b. How many A molecules are formed? _____
   c. Why are these two numbers equal?

6. a. During the time interval 0–1 s, what fraction of the A molecules react?

   b. During the time interval 10–11 s, what fraction of the A molecules react?

   c. During the time interval 24–25 s, what fraction of the A molecules react?

   d. During the time interval 40–41 s, what fraction of the A molecules react?

7. Based on the answers to CTQ 6, verify that 921 molecules of A react during the time interval 5–6 s.

8. During the time interval 100–101 s, how many molecules of A react? Explain your reasoning.

9. a. During the time interval 1–2 s, what fraction of the B molecules react?

   b. During the time interval 10–11 s, what fraction of the B molecules react?

   c. During the time interval 24–25 s, what fraction of the B molecules react?

d. During the time interval 40–41 s, what fraction of the B molecules react?

10. Based on the answers to CTQ 9, verify that 631 molecules of B react during the time interval 5–6 s.

11. During the time interval 100–101 s, how many molecules of B react? Explain your reasoning.

12. For the reaction described in Table 1:

   a. How long did it take for the reaction to come to equilibrium?

   b. Are A molecules still reacting to form B molecules at t = 500 seconds?

   c. Are B molecules still reacting to form A molecules at t = 500 seconds?

Information:

For the process in Model 2, rate of conversion of B to A = number of B molecules that react per second

$$= \frac{\Delta \text{ number of B molecules}}{\Delta t}$$

where "Δ" means "change in."

The relationship between the rate of conversion of B to A and the number of B molecules is given by equation (1):

$$\text{rate of conversion of B to A} = (\text{fraction}) \times (\text{number of B molecules}) \qquad (1)$$

where (fraction) is a specific value called a *rate constant*, k_B.

Critical Thinking Questions:

13. Rewrite equation (1), replacing "(fraction)" with "k_B."

14. What is the value of k_B in equation (1)? Be sure to include the *units* in your answer.

15. Rewrite equation (1), replacing "k_B" with the value of k_B that you determined in CTQ 14.

16. Write a mathematical equation analogous to equation (1) for the rate of conversion of A molecules into B molecules. This equation should include a constant k_A.

17. What is the value of k_A in your equation in CTQ 16? Include units.

Exercises:

1. In this activity, you calculated a rate constant for the *forward* reaction (k_A) of ________ (include units).
2. In this activity, you calculated a rate constant for the *reverse* reaction (k_B) of ________ (include units).
3. The ratio of the forward rate constant to the reverse rate constant (*i. e.,* k_A/k_B) is called the *equilibrium constant*. Calculate the equilibrium constant, K_{eq}, for the reaction in Model 2.
4. When equilibrium is reached, what fraction of the molecules in the above reaction exist as products (B)?
5. Does this equilibrium *favor* (have more of) the **products** or the **reactants** (circle one)?
6. For a reaction with $K_{eq} > 1$, does the equilibrium favor the products or the reactants?
7. For a reaction with $K_{eq} < 1$, would the equilibrium favor the products or the reactants?
8. For a reaction with $K_{eq} = 1$, would the equilibrium favor the products or the reactants?
9. Read the assigned pages in your textbook, and work the assigned problems.

Rates of Reactions

(What determines how fast a chemical reaction proceeds?)

Information:

Consider Model 1 of ChemActivity 21 concerning the equilibrium of *cis*-2-butene and *trans*-2 butene.

Critical Thinking Questions:

1. In what way(s) are the collisions between molecules at 400°C different from the collisions at room temperature?

2. What condition(s) must be satisfied for a molecular collision to result in a chemical reaction?

Model 2: Energy diagrams for three model reactions

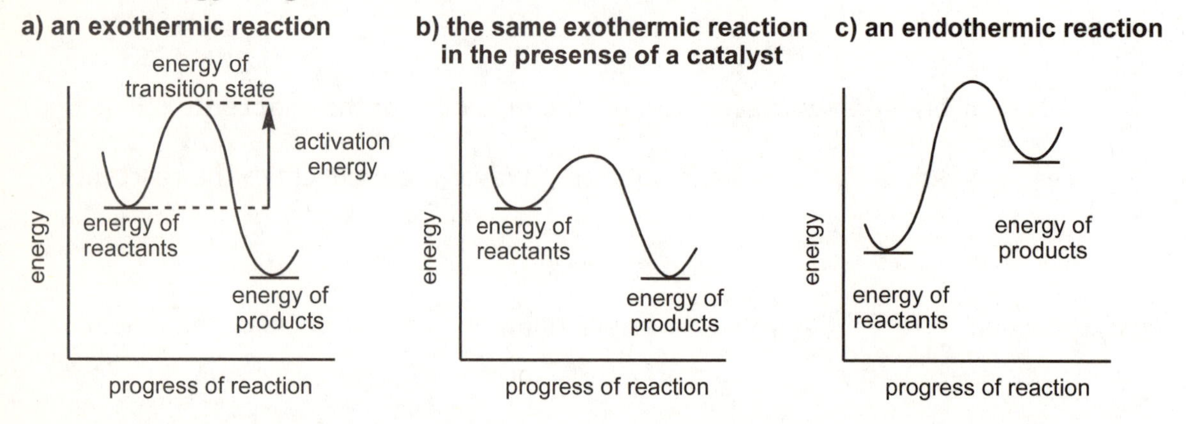

A *catalyst* increases the rate of a chemical reaction without being consumed in the reaction.

The *transition state* is the particular arrangement of atoms in the reactants at the maximum energy level.

Critical Thinking Questions:

3. What is *activation energy*?

4. Label the transition state energy in Model 2, parts (b) and (c).

5. Draw vertical arrows onto Models 2 (b) and (c) that represent the magnitude of the activation energy for conversion of the reactants to the products.

6. What happens to reactant molecules that have **less** energy than the energy of the transition state?

7. How does the addition of a catalyst affect the magnitude of the activation energy of a chemical reaction?

8. What is the role of activation energy in affecting the rate of a chemical reaction?

9. On average, do molecules have more energy at **higher** or **lower** temperatures (circle one)?

10. Why do reactions proceed faster at higher temperatures?

11. Complete the sentence:

    A catalyst increases the rate of a chemical reaction by

12. Considering Model 2, does a catalyst affect the energies of the reactants? _____

13. Considering Model 2, does a catalyst affect the energies of the products? _____

Information:

The lower the energy of a species, the more favorable it is to be formed. In exothermic reactions, the products have less energy than the reactants, and so these reactions favor conversion of most of the reactants to products. The exact equilibrium amounts of reactants and products are determined by their relative energies.

Critical Thinking Questions:

14. Considering your answer to CTQs 12 and 13, does a catalyst affect the equilibrium amounts of reactants or products? Why or why not?

15. In a chemical equilibrium, what is it that is equal?

16. The state of equilibrium is often called a dynamic equilibrium. Explain the meaning of the term *dynamic* in this context.

Exercises:

1. Most exothermic reactions are considered *spontaneous*, meaning that the products are more stable (have lower energy) than the reactants. However, *spontaneous* does not necessarily equate to *fast*. Explain why a reaction that is considered spontaneous may nevertheless not show any observable reaction.

2. Sketch a plot of energy (y axis) versus reaction progress (x axis) for a typical *exothermic* reaction. Then, draw another line onto the *same* plot which shows how the curve changes when the reaction occurs in the presence of a catalyst. Label each line. Explain how a catalyst increases the rate of a chemical reaction in terms of the meaning of your plot.

3. Consider the reaction in Exercise 2:

   a. Would this reaction *favor* formation of mostly **products** or mostly **reactants**?

   b. Would the equilibrium constant, K_{eq}, for this reaction be **less than**, **equal to**, or **greater than** 1? Circle one, and explain your choice.

   c. When this reaction had reached equilibrium, would the forward rate be **less than**, **equal to**, or **greater than** the reverse rate? Circle one, and explain your choice.

4. The burning of a piece of paper in air (to produce CO_2 and H_2O vapor) is a spontaneous exothermic reaction.

   a. What would be necessary to start this reaction?

   b. After the reaction starts, what provides the activation energy for the reaction to continue?

5. Read the assigned pages in your textbook and work the assigned problems.

Gases

(Do all gases behave the same?)

Model 1: Representation of the molecular view of water in a closed container at 20°C (293 K).

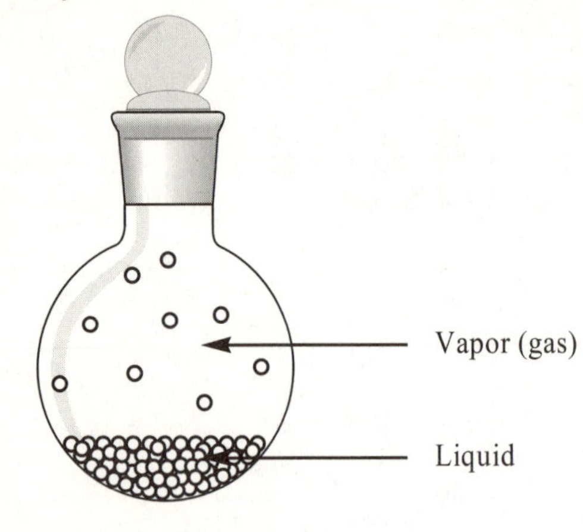

Key: ○ = 1 × 10^{21} molecules of water.

Critical Thinking Question:

1. Considering Model 1, what are the main differences between a gas and a liquid at the same temperature?

Model 2: Representation of some gases and their properties under different conditions.

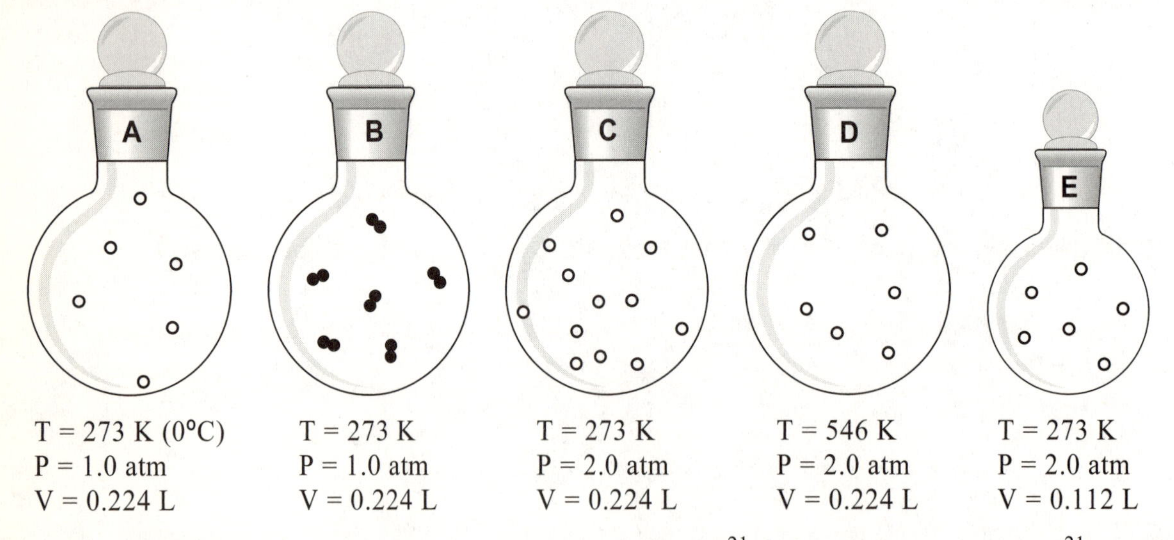

A	B	C	D	E
T = 273 K (0°C)	T = 273 K	T = 273 K	T = 546 K	T = 273 K
P = 1.0 atm	P = 1.0 atm	P = 2.0 atm	P = 2.0 atm	P = 2.0 atm
V = 0.224 L	V = 0.224 L	V = 0.224 L	V = 0.224 L	V = 0.112 L

Key: T = temperature; P = Pressure; V = volume; ●● = 1.0×10^{21} molecules N_2; ○ = 1.0×10^{21} atoms He

The standard air pressure at sea level (about 15 lb/in^2) is <u>1 atmosphere</u> (atm)

1 atm = 760 mm Hg = 760 torr

Critical Thinking Questions:

2. Compare containers A and B in Model 2. How do the properties of temperature (T) and pressure (P) change when the identity of the gas in the container is changed at constant volume?

3. Compare containers A and C in Model 2. How does the pressure of a gas change when the number of **molecules** in the container is doubled at constant volume?

4. How does the pressure of a gas change when the number of **moles** of gas in the container is doubled at constant volume?

5. Considering Model 2, how does the pressure of a gas change when the temperature is doubled at constant volume?

6. Considering Model 2, how does the pressure of a gas change when the volume of the container is doubled at constant temperature?

7. The symbol $\propto$ means "is proportional to," and the symbol "n" means "number of **moles** of gas." Based on your answer to CTQ 4, circle the correct expression.

$$P \propto n \qquad P \propto \frac{1}{n}$$

8. Based on your answer to CTQ 5, circle the correct expression.

$$P \propto T \qquad P \propto \frac{1}{T}$$

9. Based on your answer to CTQ 6, circle the correct expression.

$$P \propto V \qquad P \propto \frac{1}{V}$$

10. Add each of the three terms from the right-hand side of the expressions that you circled in CTQs 7, 8, and 9 to the expression below (all multiplied together).

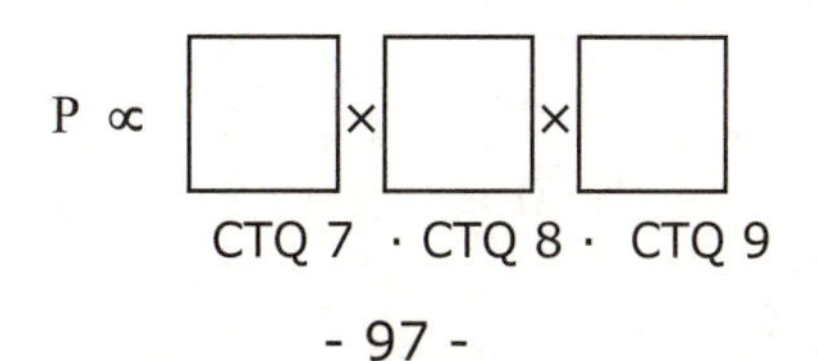

11. Since $\propto$ means "proportional to" something, it also means "equals a constant multiplied by" something. Rewrite the equation from CTQ 10 and replace the "$\propto$" with " = (constant) times".

12. The proportionality constant implied in CTQ 11 is called the **ideal gas constant**, and is given the symbol **R**. Rewrite the expression, replacing " = (constant) times" with "= R".

13. Use algebra to rearrange the equation from CTQ 12 so that all the variables are in the numerator. This equation is called the **ideal gas law.**

14. Using the number of molecules in container A in Model 2, calculate the number of moles (n).

15. Use the values of P, T, n, and V from container A in Model 2 to calculate R. Give the correct units.

16. Repeat CTQ 15 using the values of P, T, n, and V from container C, D, or E in Model 2 (your choice).

17. Compare your answers to CTQs 15 and 16. Are they the same? Why or why not?

Model C : Mixtures of gases

In a mixture of gases, the total pressure, P_T, is the sum of the pressures of the individual gases. The individual pressures are called **partial pressures**.

$$P_T = \sum_i P_i \text{, meaning } P_T = P_1 + P_2 + P_3 + \cdots \qquad (1)$$

The pressure of each gas in the mixture obeys the ideal gas law:

$$P_i = n_i \frac{RT}{V} \qquad (2)$$

Critical Thinking Questions:

18. Is equation (2) in agreement with the ideal gas law? Explain.

19. A student makes the following statement: "The ratio of the partial pressures of two gases in a mixture is the same as the ratio of the number of moles of the two gases." Is the student correct? Explain your answer.

Exercises:

1. Calculate the volume of 359 g of ethane (C_2H_6) at 0.658 atm and 75°C. (Watch the units!)

2. A 10.00-L SCUBA tank holds 18.0 moles of O_2 and 12.0 moles of He at 289 K. What is the partial pressure of O_2? What is the partial pressure of He? What is the total pressure?

3. Complete ChemWorksheet 2, Gases: Practice Worksheet.

4. Read the assigned pages in your text, and work the assigned problems.

Solutions and Molarity

(When is it dissolved, and when is it suspended?)

Suggested demonstration: Solutions, colloids, and suspensions

Information:

When two substances are mixed together and the mixture is homogeneous, we call the mixture a **solution**. The component of a solution in the greater amount is the **solvent**, and the component in the lesser amount is the **solute**. The most common solvent on earth is water. Solutes that dissolve in water may be solids (*e. g.*, salt or sugar), liquids (*e. g.*, alcohol) or gases (*e. g.*, ammonia or oxygen).

Each solute has a particular **solubility** in water—often reported as the maximum number of grams of the solute that will dissolve in 100 grams of water. A solution containing the maximum amount of dissolved solute is called a **saturated solution**.

If the solute particles are large enough to be seen (they scatter light, making the mixture *cloudy*) but not to settle out, the mixture is called a **colloid**. If the particles are larger still and can settle out over time, the mixture is called a **suspension**.

Critical Thinking Question:

1. Label each of the following as a solution, colloid, or suspension.

   a. tomato juice

   b. fog

   c. apple juice

   d. Italian salad dressing

   e. tea

   f. muddy water

   g. homogenized milk

   h. cola

2. In each of the following mixtures, name the solvent and at least one solute.

Mixture	Solvent	Solute(s)
fog		
apple juice		
cola		
NaCl(aq)		

Information:

The measure of the amount of solute dissolved in a specified amount of solution is called the **concentration** of the solution. The most common measure of concentration is **molarity**.

Molarity (M) – the moles of **solute** per liter of **solution**: $M = \dfrac{moles\ of\ solute}{L\ of\ solution} = \dfrac{mol}{L}$

For example, battery acid is approximately 6 M H_2SO_4. This means that there are 6 moles of sulfuric acid in every liter of solution. We can write two conversion factors:

$$\frac{6\ \text{moles}\ H_2SO_4}{1\ \text{L}} \quad \text{and} \quad \frac{1\ \text{L}}{6\ \text{moles}\ H_2SO_4}$$

Note how we assume—but didn't actually *write*—the word "solution" with the "1 L."

Suggested Demonstration: How to make a solution of a particular molarity

Model 1: How to make 100 mL of a 2.0-molar (2.0 M) aqueous solution of a solid

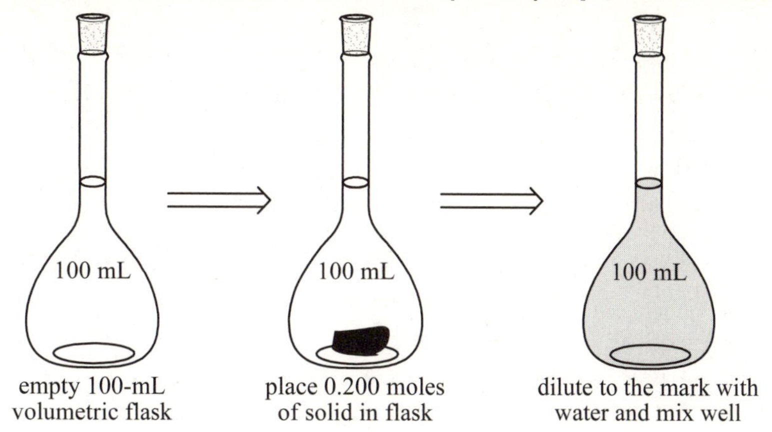

Note: volumetric flasks come in standard sizes, such as 25 mL, 50 mL, 100 mL, 250 mL, 500 mL, 1 L, *etc.*

Critical Thinking Questions:

3. In Model 1, only two-tenths of a mole of a solid was used to make a 2.0-molar solution. Explain how this is possible.

4. An experiment calls for 50 mL of a 0.50 M aqueous solution of sodium hydrogen carbonate (sodium bicarbonate, or baking soda). *Describe* (with amounts) the steps you would use in order to make up such a solution, such that you have none left over.

Information:

When **diluting** a solution, the moles of the solute do not change. Therefore,

$$\text{moles (before)} = \text{moles (after)} \qquad \textit{or} \qquad \text{mol}_1 = \text{mol}_2$$

Since M= mol/L, (M × L) is equal to moles. So we can write:

$$\text{M} \times \text{L (before)} = \text{M} \times \text{L (after)}$$

or

$$M_1V_1 = M_2V_2 \text{ (where V = volume)} \qquad (1)$$

This is an equation that can be used for **dilutions**. It may also be rearranged to give:

$$M_2 = \frac{M_1V_1}{V_2} \qquad (2)$$

Model 2: How to make 100 mL of 0.060 M $CuSO_4$ from 0.60 M $CuSO_4$

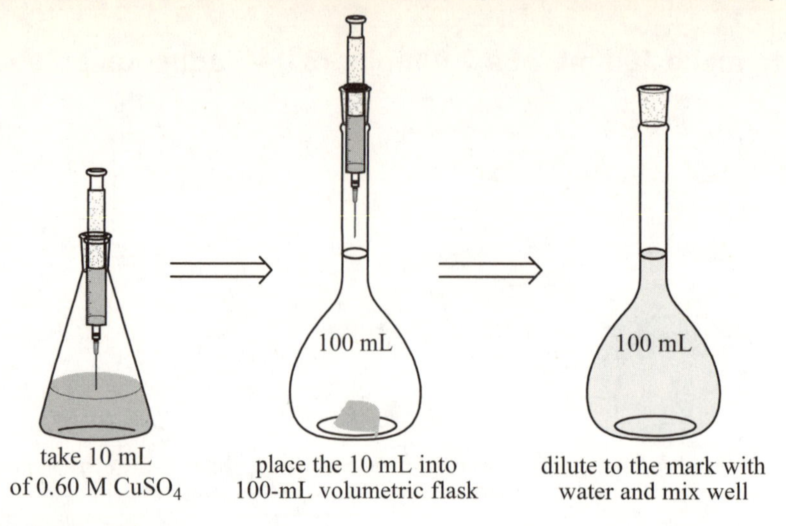

Critical Thinking Questions:

5. Consider Model 2. Verify that diluting 10 mL of 0.60 M copper(II) sulfate to 100 mL will produce a 0.060 M solution. (Identify M_1, V_1, and V_2 and calculate M_2).

6. How many mL of 0.60 M copper(II) sulfate would need to be added to the flask and diluted to a total volume of 100 mL to make an 0.1 M solution? (Hint: Which variable in equation (1) is not known?)

Exercises:

1. If 50 mL of concentrated (18 M) sulfuric acid is diluted to a total volume of 1.0 L, what is the new concentration?

2. Iodine dissolves in various organic solvents, such as dichloromethane (CH_2Cl_2), in which it forms an orange solution. What is the molarity of I_2 when 5.00 g iodine is dissolved in enough dichloromethane to make 50.0 mL of solution?

3. A patient's blood calcium level is 9.2 mg/dL. What is this concentration in molarity? Make a unit plan first.

4. Read the assigned pages in your textbook, and work the assigned problems.

Hypotonic and Hypertonic Solutions

(Is it more concentrated or more dilute?)

Information: Units of concentration

The *concentration* of a solution is a measure of the amount of solute dissolved in a specified amount of solution. A solution that is more *concentrated* has more solute per unit of volume than one that is more *dilute*.

We saw in ChemActivity 24 that the most common measure of concentration is molarity.

Molarity (M) – the moles of solute per liter of solution: $M = \dfrac{moles\,of\,solute}{L\,of\,solution} = \dfrac{mol}{L}$

The other common measures of concentration are not done using moles, but using either mass (weight) or volume. These are commonly reported in percent. "Percent by weight" and "percent by volume" are common terms, and may be represented as %(w:w) or %(v:v). Weights (or masses) are in grams, and volumes in milliliters. The first letter in the parenthesis represents the units of the **solute**, and the second is for the **solution**.

Since *percent* means *"per hundred,"* a 3.2%(v:v) aqueous solution of alcohol would mean 3.2 mL of alcohol are in every 100 mL of solution. We can write this as a conversion factor:

$$\frac{3.2\text{ mL alcohol}}{100\text{ mL}} \quad \text{and} \quad \frac{100\text{ mL}}{3.2\text{ mL alcohol}}$$

Note how we assume—but didn't actually *write*—the word "solution" with the "100 mL."

Critical Thinking Questions:

1. Write the two conversion factors for each of the following concentrations.

   a. 10%(v:v) acetone in water

   b. 0.90%(m:v) NaCl(aq)

   c. 5.0%(w:w) $NaHCO_3$(aq)

   d. 5.0%(w:v) $NaHCO_3$(aq)

2. The solutions in CTQ 1 (c) and (d) are often considered to be the same concentration even though the units are different. Explain how this can be. (Hint: What is the density of water?)

3. Sometimes when concentrations are reported for aqueous solutions of solids, they do not say whether they are by mass or by volume. Explain why it is probably okay to assume they are %(m:v).

4. Percent means parts per hundred. We can also report concentrations in parts per thousand (ppt), parts per million (ppm), or parts per billion (ppb). The EPA safe limit for lead in drinking water is 15 ppb. Write the two conversion factors for this concentration.

Information: Hypotonic and hypertonic solutions

Osmotic pressure (π) is the pressure that water exerts against a semipermeable membrane, such as a cell membrane. *The* ***higher*** *the* ***solute*** *concentration, the* ***lower*** *the osmotic pressure.* Red blood cells have an osmotic pressure equal to that of a 0.90% (m:v) solution of sodium chloride. In this solution, since there are 2 moles of ions per mole of NaCl, the concentration of **ions** is twice as much, or 1.8% (m:v).

A solution that has a higher ion concentration than this is called **hypertonic**; one with a lower ion concentration is called **hypotonic**. It is important when giving fluids intravenously that they be isotonic, to avoid hemolysis (bursting) or crenation (shriveling) of red blood cells.

Critical Thinking Questions:

5. Given the definitions of the terms *hypotonic* and *hypertonic*, what do you suppose an *isotonic* solution would be?

6. Is a 1.0 M solution of NaCl hypotonic, hypertonic, or isotonic to red blood cells? (Hint: Convert to % m:v—make a unit plan!)

7. An experiment calls for 10.0 mL of a 4.00% aqueous solution of sodium tetraborate ("borax"). Describe how you would make up such a solution so that you will not have any left over.

Information:

Electrolytes in blood are often measured in equivalents per liter (Eq/L). An *equivalent* of an ion is the amount of that ion that gives 1 mole of positive or negative charge. This gives rise to conversion factors such as:

$1\ mol\ Na^+ = 1\ Eq\ Na^+$ $1\ mol\ Ca^{2+} = 2\ Eq\ Ca^{2+}$ $1\ mol\ SO_4^{2-} = 2\ Eq\ SO_4^{2-}$

Critical Thinking Questions:

8. A patient has a blood calcium level of 4.6 mEq/L.
   a. Write the two conversion factors for converting moles of calcium into equivalents of calcium.

   b. Write a **unit plan** to convert mEq/L into mol/L.

   c. What is the patient's blood calcium level in molarity?

Exercises:

1. What is the blood calcium level of the patient in CTQ 8 in mg/dL? (Unit plan!)

2. If the lead level in some drinking water is 15 ppb, how many mL of the water would a person have to drink in order to ingest one gram of lead? (Unit plan!)

3. Is a 0.9 M $CaCl_2$ solution isotonic to red blood cells? Explain why or why not, without doing a calculation.

4. Read the assigned pages in your textbook, and work the assigned problems.

Acids and Bases

(What happens when hydrogen ions are transferred between species?)

Information: Two definitions of acids and bases

Arrhenius definitions

An acid is a species that dissociates into H^+ (hydrogen) ions and anions when dissolved in water.

A base is a species that dissociates into OH^- (hydroxide) ions and cations when dissolved in water.

Brønsted-Lowry definitions

An acid donates a proton (H^+ ion) to another species.

A base accepts a proton (H^+ ion) from another species.

The Brønsted-Lowry definition explains why the hydrogen ions (H^+) in water are actually hydronium ions—a water molecule (H_2O) has accepted the proton to become hydronium (H_3O^+).

Table 1: Some common acids and bases

Type of electrolyte	Acids	Bases
Strong (all not listed here are weak)	HCl HBr HI H_2SO_4 HNO_3 $HClO_4$	LiOH NaOH KOH $Ca(OH)_2$ $Sr(OH)_2$ $Ba(OH)_2$
Weak	$HC_2H_3O_2$ HCN	$Mg(OH)_2$

Recall that when strong acids and bases dissolve in water, they dissociate completely into ions, while weak acids and bases dissociate only slightly.

Critical Thinking Questions:

1. Write the three species that actually exist in significant amounts in a one-tenth molar aqueous solution of HCl.

2. Explain why hydrogen ion (H^+) is *not* one of the three species in CTQ 1.

3. Write the three species that actually exist in significant amounts in a one-tenth molar aqueous solution of LiOH.

Information: Hydronium-hydroxide balance

In pure water, a small amount of self ionization occurs, with one water molecule acting as an acid (donating a proton) and another as a base (accepting a proton):

$$H_2O(l) + H_2O(l) \text{ qwe } H_3O^+(aq) + OH^-(aq)$$

In pure water, the concentrations of hydronium ion and hydroxide ion are each 1.0×10^{-7} M. Furthermore, the product of the concentrations of hydronium ion and hydroxide ion in aqueous solution at 25°C is always 1.0×10^{-14} M.

$$[H_3O^+][OH^-] = 1.0\times10^{-14}$$

This means that if the hydronium ion concentration $[H_3O^+]$ increases, the hydroxide ion concentration $[OH^-]$ decreases. This relationship allows us to calculate the amounts of hydronium ion and hydroxide ion in any solution of strong acid or base.

Critical Thinking Questions:

4. What is the hydronium ion concentration in a 1.0×10^{-5} M aqueous solution of HCl?

5. What is the hydroxide ion concentration in an aqueous solution if the hydronium ion concentration is 1.0×10^{-5} M?

Information: pH

pH (the "power of hydrogen") is defined as the negative of the logarithm of the molar concentration of hydronium ions (without units):

$$pH = -\log[H_3O^+]$$

Therefore, in pure water, the pH is $-\log(1.0\times10^{-7})$, which equals 7. pH values below 7 are called *acidic*; those above 7 are termed *basic* or *alkaline*. pH values can actually go below 0 and above 14, though this is not commonly seen.

Table 2: The relationship between acidity, pH, and the hydronium and hydroxide ion concentrations of a solution.

Relative Concentrations	pH	Solution
$[H_3O^+] > [OH^-]$	< 7	acidic
$[H_3O^+] < [OH^-]$	> 7	basic
$[H_3O^+] = [OH^-]$	$= 7$	neutral

Critical Thinking Questions:

6. What is the pH of the 1.0×10^{-5} M aqueous solution of HCl from CTQ 3? (Be sure you can enter this into your calculator correctly, *e. g.*, `1 . 0 EE 5 ± log ±`)

7. According to Table 2, which ion does the solution in CTQ 5 contain more of:

   hydronium or **hydroxide** (circle one)?

8. Does your answer to CTQ 7 agree with your answers to CTQs 4 and 5?

9. a. If the pH of a cola drink is 3.2, what is the hydronium ion concentration? Be sure you can enter this into your calculator correctly, *e. g.*, `3 . 2 ± 10ˣ` (the `10ˣ` key is often an `inverse` or 2nd `log`).

   b. What is the hydroxide ion concentration in the cola?

10. a. What is the hydroxide ion concentration in a 1.0×10^{-5} M aqueous solution of **NaOH**?

    b. What is the pH of this solution? (Careful!)

Model: Conjugate acid-base pairs

According to the Brønsted-Lowry theory, a reaction of an acid and a base involves a proton (*i. e.*, hydrogen ion) transfer from the acid to the base. Two ions or molecules that *differ only by that one hydrogen ion* make up a **conjugate acid-base pair**. Three example pairs are shown below:

H_3O^+ and H_2O $\quad$ H_2O and OH^- $\quad$ NH_4^+ and NH_3

The **conjugate acids** have one more proton (H^+) than the **conjugate bases**. For example, consider the reaction below:

conjugate acid-base pair (CH₃COOH and CH₃COO⁻)

$$CH_3COOH + H_2O \rightleftharpoons CH_3COO^- + H_3O^+$$

acetic acid + water $\quad$ acetate + hydronium ion

conjugate acid-base pair (H₂O and H₃O⁺)

Sometimes we simplify the naming by saying an acid and a base react to give a conjugate acid and conjugate base.

$$\underset{\text{acid}}{CH_3COOH} + \underset{\text{base}}{H_2O} \rightleftharpoons \underset{\text{conjugate base}}{CH_3COO^-} + \underset{\text{conjugate acid}}{H_3O^+}$$

- The conjugate base is what results after the acid gives up a hydrogen ion; so we say that *acetate is the conjugate base of acetic acid*.
- The conjugate acid is what results after the base picks up a hydrogen ion; so we say that *hydronium ion is the conjugate acid of water*.

Critical Thinking Questions:

11. For the equations below, identify the acid and base on the reactant side and the conjugate acid and conjugate base on the product side. Draw a line to connect conjugate acid-base pairs together.

    a. $HCN(aq) + OH^{-}(aq)$ qwe $H_2O(l) + CN^{-}(aq)$

    b. $F^{-}(aq) + H_3O^{+}(aq)$ qwe $H_2O(aq) + HF(aq)$

    c. $NH_3(aq) + HCl(aq)$ qwe $NH_4^{+}(aq) + Cl^{-}(aq)$

Exercises:

1. Which definition of an acid—Arrhenius or Brønsted—is more complete for aqueous solutions? Explain.

2. Calculate the pH of each of the following aqueous solutions.
   a. 1.0×10^{-4} M nitric acid
   b. 5.0×10^{-3} M hydrobromic acid
   c. a 1.0×10^{-6} M solution of the diprotic acid H_2SO_4 (diprotic means that each molecule of H_2SO_4 donates two hydrogen ions to water molecules)
   d. 0.0012 M $Ca(OH)_2$
3. Calculate the hydroxide ion concentrations of the solutions in Exercise 2.

4. Write formulas for the conjugate bases of the acids in Exercises 2a and 2b.

5. Read the assigned pages in your textbook, and work the assigned problems.

Buffers

(How do acids and bases react together?)

Model 1: A buffer system

Consider a solution containing both the **weak** acid acetic acid and its **conjugate base**, sodium acetate. This sets up the equilibrium shown below. (The sodium ion is just a spectator, and does not show up in the equilibrium expression.)

CH_3COOH	+ H_2O	qwe	CH_3COO^-	+	H_3O^+
acetic acid	water		acetate		hydronium ion
(acid)	(base)		(conjugate base)		(conjugate acid)

Critical Thinking Questions:

1. Suppose some strong base (hydroxide ion) is added to the buffer system in Model 1. Circle the two species in the model that could react with the hydroxide.

2. Write a chemical equation for the reaction of hydroxide ion being neutralized by reacting with acetic acid, producing acetate and water.

3. Explain why addition of strong base (hydroxide ion) to the solution in Model 1 will not cause a great change in the pH of the solution, as long as the acetic acid is not used up.

4. Draw a <u>box</u> around the one species in Model 1 that could react with and neutralize any added hydronium ion.

5. Explain why addition of strong acid will not cause the pH of the solution in Model 1 to change much, as long as plenty of acetate is present.

Model 2: A solution of a strong acid and its conjugate base

Consider a solution containing both the **strong** acid hydrochloric acid and its **conjugate base**, sodium chloride. This sets up the system shown below.

HCl	+ H_2O	ssd	Cl^-	+	H_3O^+
hydrochloric acid	water		chloride		hydronium ion
(acid)	(base)		(conjugate base)		(conjugate acid)

Critical Thinking Questions:

6. Recalling that strong acids dissociate completely in water, draw a large 'X' through the species in Model 2 that is not present in any significant amount.
7. Why is a forward arrow used in Model 2 (ssd) instead of an equilibrium arrow (qwe)?

8. Explain the following statement: There is no species in Model 2 that can neutralize added hydronium ion.

Information: A summary

A solution containing both a **weak** acid and its **conjugate base** is resistant to changes in pH when small amounts of acid or base are added. This solution is called a **buffer**.

A solution of a **strong** acid and its conjugate base is **not** a buffer, since any added strong acid will not be neutralized and will just increase the H_3O^+ concentration. Similarly, a solution of a **strong base** and its conjugate acid is **not** a buffer.

Exercises:

1. State if each solutions would be useful as a buffer or not. Then explain the reason for your choice.
   a. A solution containing 0.08 M NaCN and 0.10 M HCN

   b. A solution containing 0.05 M NaOH in H_2O

   c. A solution containing 0.25 M HCl and 0.20 M NaCl

   d. A solution containing 0.05 M NH_4Cl and 0.10 M NH_3

   e. A solution containing 0.20 M KF and 0.15 M HF

2. Considering the solution in Exercise 1b.
   a. Write the three species that would be present in significant amounts.

   b. Is there any species present that can neutralize added hydroxide? Explain.

3. Complete ChemWorksheet 3, Stoichiometry Practice Worksheet 2.
4. Read the assigned pages in your textbook, and work the assigned problems.

Alkanes, Cycloalkanes and Alkyl Halides

(What makes a molecule "organic?")

Information:

Organic molecules are based on carbon backbone structures. Of these, hydrocarbons contain only carbon and hydrogen. The hydrocarbons containing no multiple bonds are called **alkanes**. The first ten "straight-chain" alkanes are shown in Table 1.

Table 1. Names and structures of the first ten alkanes (and alkyl groups)

Condensed Structure	Name of alkane	"Stick" Structure	Condensed Structure	Name of alkyl group substituent
CH_4	methane	none	$-CH_3$	methyl
CH_3CH_3	ethane		$-CH_2CH_3$	ethyl
$CH_3CH_2CH_3$	propane		$-CH_2CH_2CH_3$	propyl
$CH_3CH_2CH_2CH_3$	butane		$-CH_2CH_2CH_2CH_3$	butyl
$CH_3(CH_2)_3CH_3$	pentane		$-CH_2(CH_2)_3CH_3$	
$CH_3(CH_2)_4CH_3$	hexane		$-CH_2(CH_2)_4CH_3$	
$CH_3(CH_2)_5CH_3$	heptane		$-CH_2(CH_2)_5CH_3$	
$CH_3(CH_2)_6CH_3$	octane		$-CH_2(CH_2)_6CH_3$	
$CH_3(CH_2)_7CH_3$	nonane		$-CH_2(CH_2)_7CH_3$	
$CH_3(CH_2)_8CH_3$	decane		$-CH_2(CH_2)_8CH_3$	

The structures shown in Table 1 are called **condensed structures**. They can be written as a molecular formula by adding all the carbons and hydrogens (*e. g.*, ethane = C_2H_6) or expanded into a complete Lewis structure (*e. g.*, ethane, shown below):

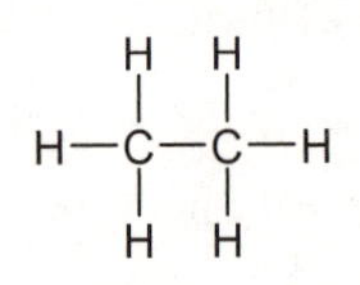

Critical Thinking Questions:

1. Draw a complete Lewis structure for propane.

2. Complete Table 1 by filling in the missing "stick" **structures** and **names** of the alkyl group substituents.

3. Using grammatically correct English sentences, describe how one can derive the name of an alkyl group substituent from the name of the corresponding alkane.

Information:

Once organic molecules get large, it is convenient to write them as "**stick structures**." Each "end" and "bend" is a carbon, and the hydrogens are not shown. Stick structures are commonly used for **cycloalkanes**—alkanes in which the carbons are connected to make a ring ("head" to "tail.") Some common cycloalkanes are shown in Table 2.

Table 2. Names and structures of some common cycloalkanes

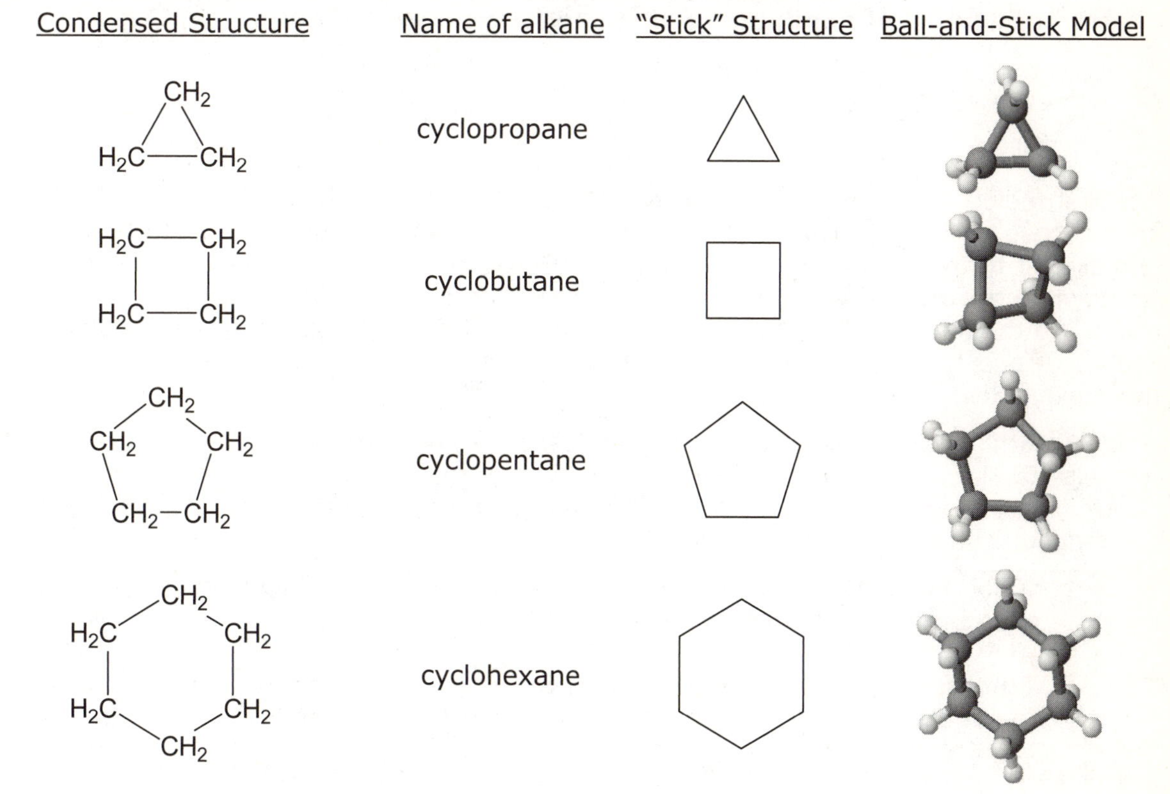

Critical Thinking Question:

4. Draw condensed and "stick" structures for cycloheptane.

Information: Rules for naming branched alkanes

Alkanes are **branched** when the carbons are not connected in a continuous manner. For example, if you cannot trace all the carbons with your pencil (or finger) from one carbon to the next without lifting your pencil or retracing a portion, there is a branch. The branches are called **alkyl groups**. Any such **side chain**, even if it is not an alkyl group, is called a **substituent**. The name of the molecule is then determined as follows:

1. Find the longest continuous chain of carbons to get the **base name**.
2. Name each **substituent** alphabetically. If there are two equivalent substituents, use the prefixes di-, tri-, tetra, *etc.* in front of them.
3. Mentally number each carbon of the base chain starting at the end closest to the branch. Finally, **give each substituent a number** that corresponds to the number of the carbon of the base chain to which it is connected.

For example, consider the two molecules in Figure 1:

Figure 1: Example of naming two different five-carbon alkanes

$CH_3CH_2CH_2CH_2CH_3$

pentane

4 3 2 1

$CH_3CH_2CH-CH_3$ (with CH_3 attached to carbon 2)

2-methylbutane

The first molecule in Figure 1 is **pentane**. The second has only four continuous carbons, and so the base name is **butane**. The carbons of the butane are numbered starting from the end closest to the $-CH_3$ group (called a **methyl group**, see Table 1 above). This numbering is shown above the molecule in the Figure. Then the number "2" is assigned to the methyl group. So the name is **2-methylbutane**. Note that there is no space between "methyl" and "butane," and there is a hyphen between the number and letter. See your textbook for more examples.

Note something else about the two molecules in Figure 1: they have the same molecular formula (C_5H_{12}). Molecules that have the same molecular formula but are arranged differently are called **isomers**. So, pentane and 2-methylbutane are isomers. We will learn more about isomers in ChemActivities 29-31.

Some common substituent names are shown in Table 3. If the substituent is a halogen, then the molecule is called a **haloalkane** or **alkyl halide**.

Table 3. Common substituent names in alkanes and haloalkanes. "R" indicates where the substituent is attached to the "Rest" of the molecule (the main chain).

Substituent	Name	Stick structure	Substituent	Name
$-CH(CH_3)CH_3$	isopropyl	[stick structure]	-F	fluoro
$-CH_2CH(CH_3)-CH_3$	isobutyl	[stick structure]	-Cl	chloro
$-CH(CH_3)CH_2CH_3$	*sec*-butyl	[stick structure]	-Br	bromo
$-C(CH_3)_2-CH_3$	*tert*-butyl	[stick structure]	-I	iodo

Critical Thinking Question:

5. Complete the following table:

Name	Lewis structure	condensed formula	stick structure
2-chloropropane			
		CH_3 $CH_2CH_2CH_3$ $CH_3CH_2CH\text{-}CH\text{-}CH\text{-}CH_3$ CH_2CH_3	
isopropylcyclohexane			
			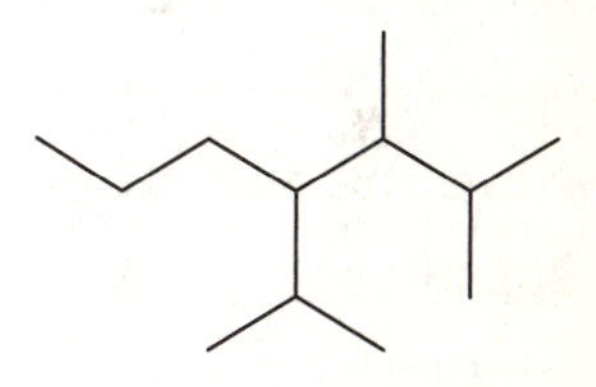

Exercises:

1. Draw (any style) and name all 5 of the differently branched isomers of C_6H_{14}.

2. Complete ChemWorksheet 4: Functional Groups.
3. Read the assigned pages in your textbook and work the assigned problems.

Conformers

(How and why do molecules "twist?")

Model 1: Representations of ethane in its most favorable conformation

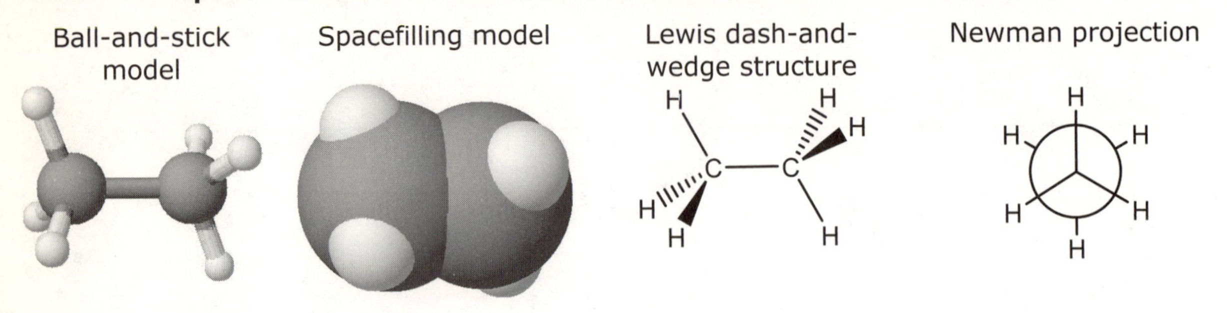

Critical Thinking Questions:

1. Each representation in Figure 1 shows ethane (CH_3CH_3) in its lowest potential energy (most favorable) conformation. If you have a model set available, make a model of ethane and rotate the single bonds until it is in this conformation.

   a. Construct an explanation for why this conformation is the most favorable.

   b. Consider the Newman Projection in Figure 1. What atom is at the center of the picture?

   c. The atom you named in CTQ 1b is represented as a large circle or disc. What single atom is hidden from view behind the disc?

2. Consider the Newman projection shown below of ethane in (nearly) its least favorable conformation. If you have a model set available, rotate the single bonds until it is in this conformation. Draw a dash-and-wedge structure for this conformation.

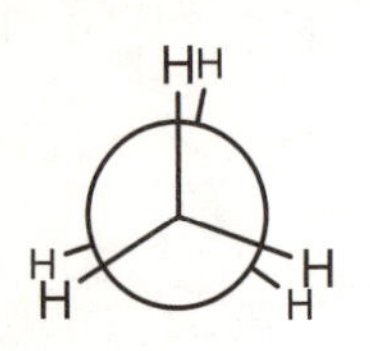

3. In your own words, explain what the term *conformation* means, as applied to ethane.

4. Structures that represent the same molecule in different conformations are termed conformers. Construct an explanation for why the conformer in Model 1 is called **staggered** and the conformer in CTQ 2 is called **eclipsed**.

5. Look at your model "end-on" and compare with the Newman projections below. The angle you observe between the hydrogens at the top of each structure in **boldface** is called the torsional angle. Circle the correct torsional angle for each conformer.

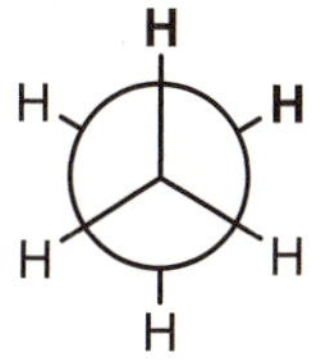

staggered

Torsional angle (staggered)
0°
60°
120°
180°

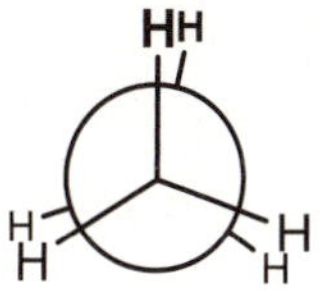

eclipsed

Torsional angle (eclipsed)
0°
60°
120°
180°

6. What repulsive forces will cause ethane in the eclipsed conformation to quickly adopt the staggered conformation?

Model 2: Some representations of alkanes with 2, 3 and 4 carbons

Skeletal or "Stick" structure	Ball-and-stick structure	Lewis structure	Dash-and-wedge structure

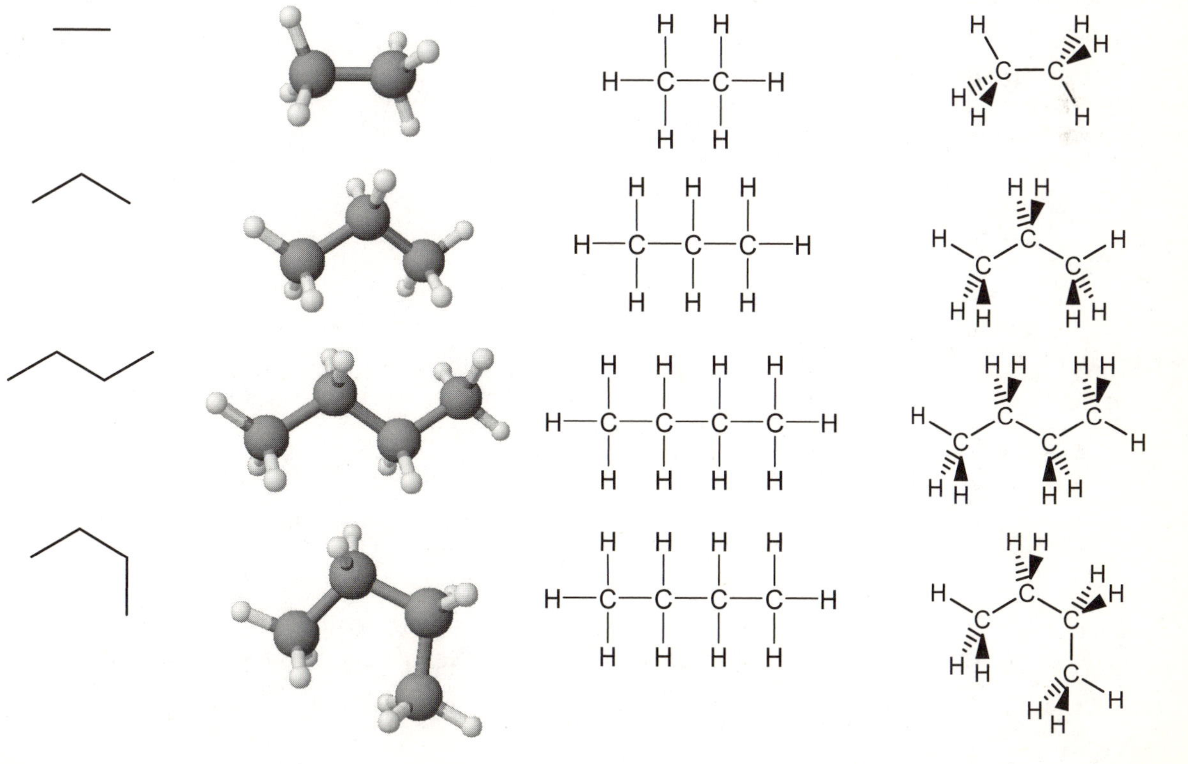

Critical Thinking Questions:

7. Underneath each stick structure in Model 2, write the name of the alkane.

8. The ball-and-stick structure below matches one of the two butane conformations shown in Model 2, looking "end-on" down the bond between carbons 2 and 3. Circle the stick structure in Model 2 that matches the structure below.

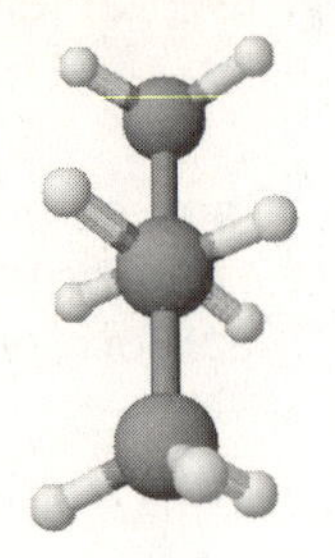

9. Complete the Newman projection at the right in CTQ 7 by adding H or CH_3 groups so that it represents the conformation shown in the ball-and-stick structure at the left.

10. Consider the other stick structure for butane in Model 2 (the one you did not circle in CTQ 7), sighting down the C2–C3 bond.

    a. Is this structure **staggered** or **eclipsed** (circle one)? You may use your model to help you.

    b. Draw a Newman projection for this conformation of butane.

    c. At room temperature, the single bonds in butane are continually rotating through the staggered and eclipsed conformations. However, the molecule spends more time closer to one of the two extremes. Which conformation is it likely to spend more time in—

    staggered or **eclipsed** (circle one)?

Model 3: Some conformations of pentane

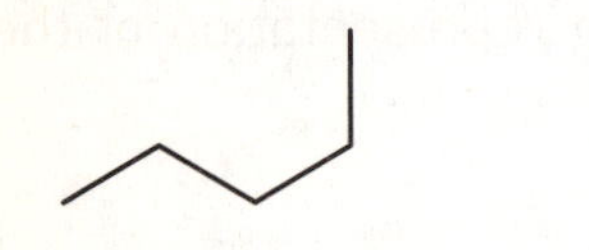

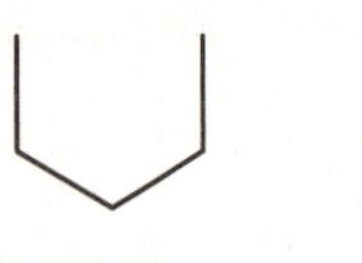

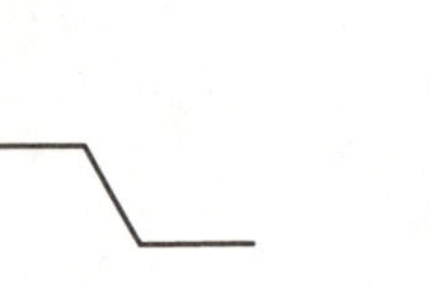

Critical Thinking Questions:

11. Circle the two structures of pentane in Model 3 that are in their most favorable conformation. Explain why these are the most favorable.

12. Draw a line to connect the two structures that you circled in Model 3. These two structures are in the same conformation, but the molecule as a whole is rotated.

13. Find two other structures in Model 3 that are identical (*i. e.,* in the same conformation), and draw a second line to connect them.
14. What is the total number of distinct conformers of pentane shown in Model 3?

Exercises:

1. Draw a wedge and dash-bond representation of pentane in its most favorable conformation.

2. Consider the molecule 2-methylbutane.
   a. Using the templates at the right, complete two staggered Newman projections for 2-methylbutane: one sighting down the C1–C2 bond and the second sighting down the C2–C3 bond.

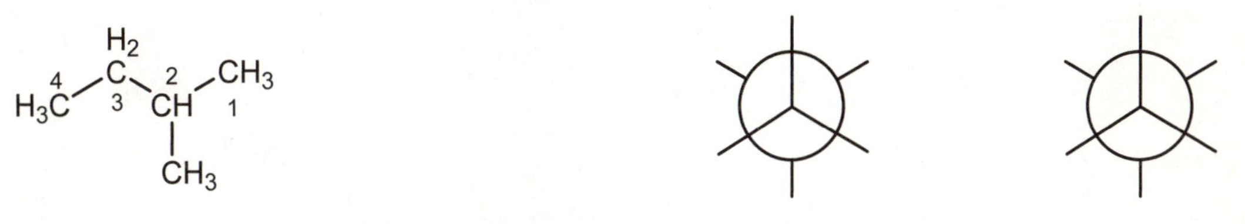

   b. Complete the eclipsed Newman projection for 2-methylbutane sighting down the C2–C3 bond.

3. Sugars and other complex molecules are often depicted using a representation called a Fisher projection. In a Fisher projection all <u>horizontal</u> bonds are assumed to come out of the page toward you (wedge bonds) and all <u>vertical</u> bonds are assumed to go back into the page, away from you (dash bonds). Draw a wedge and dash representation of the Fisher projection of glyceraldehyde shown below.

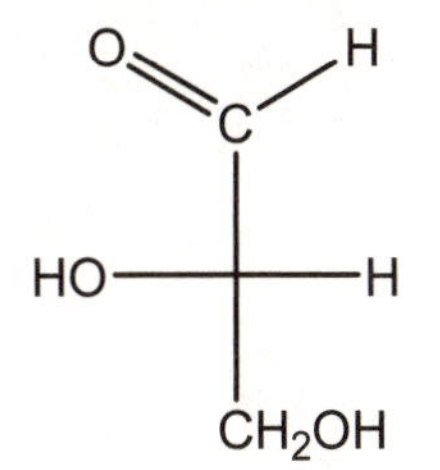

4. Read the assigned sections in your text and work the assigned problems.

Constitutional and Geometric Isomers

(Are they identical, or are they isomers?)

Model 1: Representations of some organic molecules

Skeletal or "Stick" structure	Lewis structure	Ball-and-stick structure

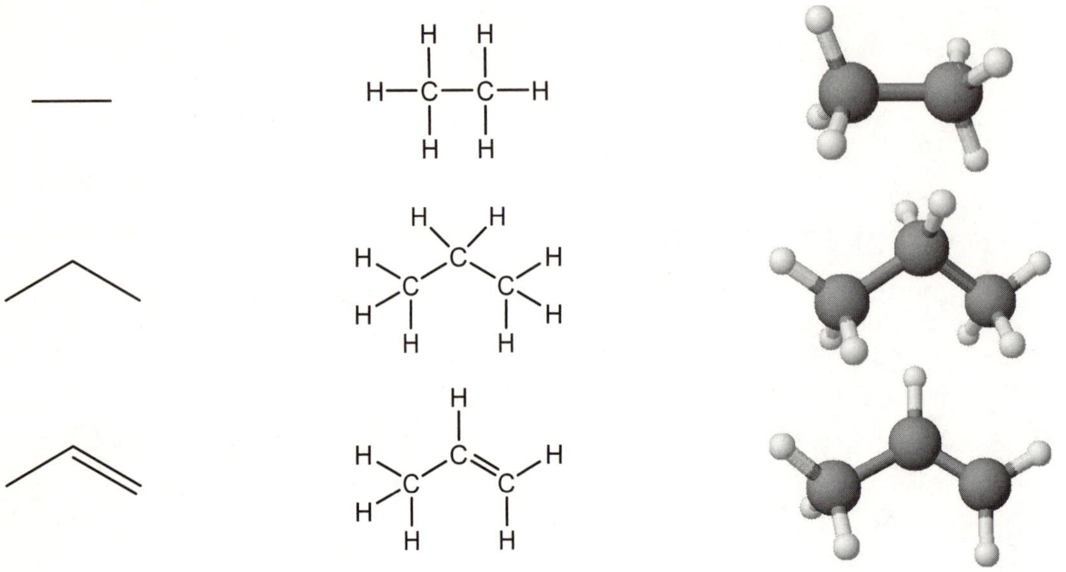

Critical Thinking Questions:

1. Consider the Lewis structures in Model 1. How many covalent bonds does each carbon have?

2. In skeletal representations, the hydrogens are not shown. Explain how it is still possible to tell how many hydrogens there are on each carbon.

3. Draw a Lewis structure representation of the molecule for which a skeletal representation is shown below.

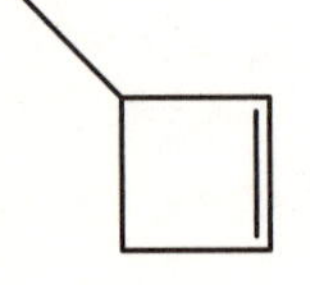

Model 2: Constitutional Isomers

Column 1		Column 2	
structure	molecular formula	structure	molecular formula
	C_5H_{12}		
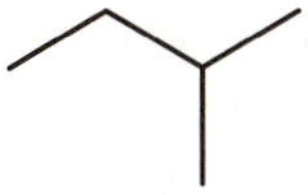			
		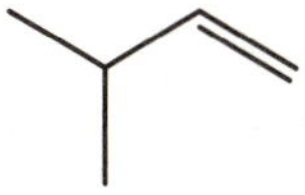	

Critical Thinking Questions:

4. Complete Model 2 by writing in the missing molecular formulas in both columns.

5. What do the molecules in a given column (1 or 2 in Model 2) have in common with the other molecules in that column?

6. What do the molecules in a given column **not** have in common with the other molecules in that column?

7. All the structures in a given column are constitutional isomers of one another, but the structures in Column 1 are not constitutional isomers of structures in Column 2. Based on this information, write a definition for the term constitutional isomers.

8. If the molecule shown below were placed into Model 2, would it belong in **Column 1** or **Column 2** (circle one)? Explain your choice.

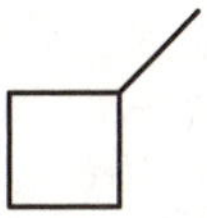

Model 3: Representations of methylcyclobutane

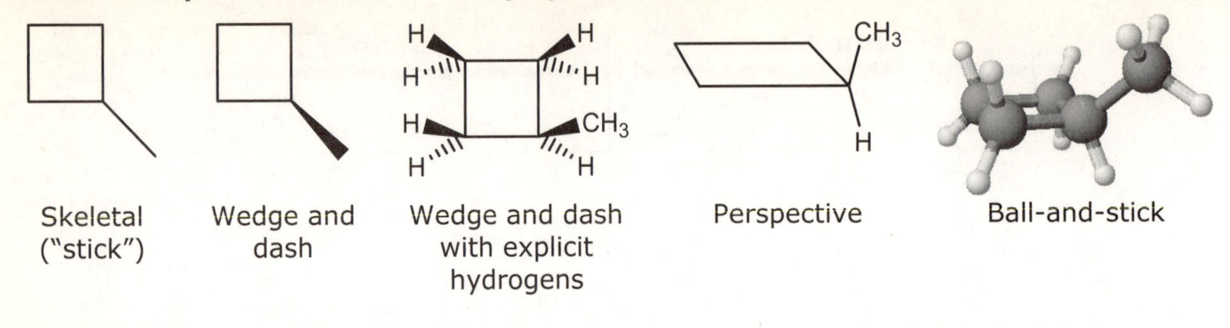

Model 4: 1,2-dimethylcyclobutane, shown with ring carbons numbered 1–4

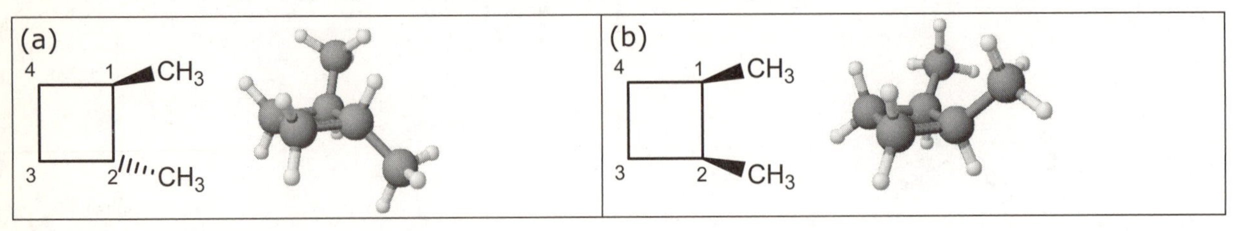

Critical Thinking Questions:

9. Are the molecules in boxes (a) and (b) of Model 4 constitutional isomers of each other? Explain.

10. *Other than bonds to carbons within the ring*, what two groups are bonded to the following carbons?
    a. carbon 1 in box (a)?
    b. carbon 1 in box (b)?
    c. carbon 2 in box (a)?
    d. carbon 2 in box (b)?

11. If you have access to a model kit, make models of the two molecules in Model 4 (C = black; H = white; use the short bonds for single bonds). Is it possible to rotate single bonds in the models such that the molecule in box (a) is the same as the one box (b)?

Information:

Since each carbon in the molecule in box (a) in Model 4 is bonded to the same four groups as the corresponding carbon in the molecule in box (b), the molecules are said to have the same **connectivity**.

You confirmed in CTQ 11 that the two structures of 1,2-dimethylcyclobutane shown above are not simply **conformers** of each other.

Imagine that the four carbons of the cyclobutane ring define a plane. In one structure, the two methyl groups are on the *same side* of this plane, and in the other they are on *opposite sides* of the plane. The single bonds in the ring cannot rotate without breaking the ring. Two groups on the *same side* of the plane are considered to be *cis* to one another. Groups on opposite sides are called *trans*.

Geometric isomers (*cis-trans* isomers) are molecules that have the same connectivity and differ only in the geometric arrangement of groups.

Critical Thinking Questions:

12. Label one box in Model 4 with the name "*cis*-1,2-dimethylcyclobutane" and the other with the name "*trans*-1,2-dimethylcyclobutane." Then add perspective representations into each box.

13. Draw wedge-and-dash and perspective representations of *cis*- and *trans*-1,3-dimethylcyclobutane. (Note: that is "1,3-dimethyl," not "1,2-dimethyl.")

Exercises:

1. Indicate if the following pairs of structures are *identical*, *conformers*, *geometric isomers*, *constitutional isomers*, or *not isomers*.

   a. 1,2-dimethylcyclobutane and 1,3-dimethylcyclobutane

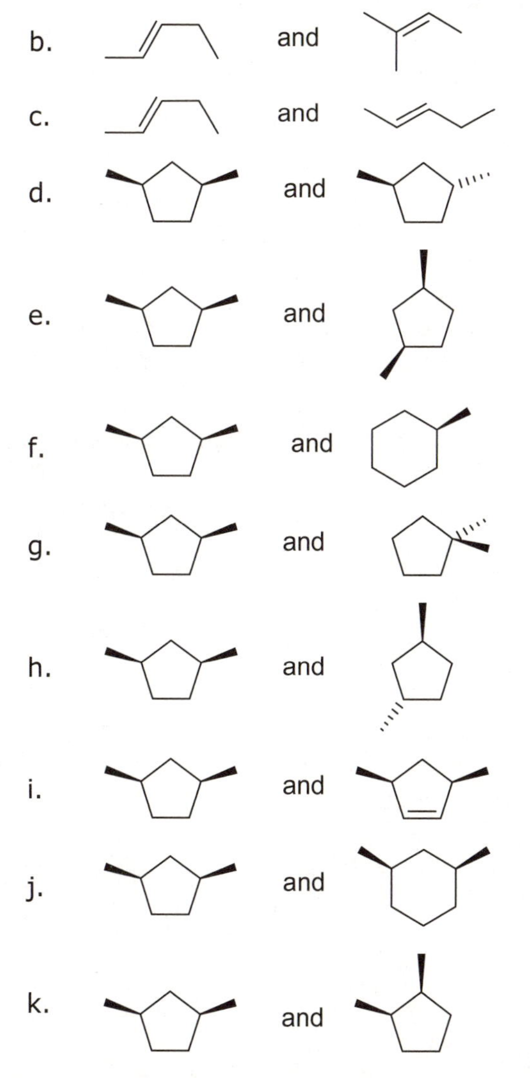

2. Draw a structure for a molecule not shown in this activity that would belong in Column 2 of Model 2.

3. Read the assigned pages in your textbook and work the assigned problems.

Isomers

(What are some different types of isomers?)

Information: Review of some types of isomers, from least to most similar

1. **Constitutional isomers**: molecules with the same molecular formula but different structures (different connectivity).
2. **Geometric isomers** (*cis-trans* isomers): molecules that have the same connectivity and differ only in the geometric arrangement of groups.
3. **Conformational isomers** (conformers): molecules that can be interconverted by rotation around single bonds.

Model 1: Alkenes

As we saw in ChemActivity 21, there is no free rotation around double bonds at room temperature. This means the two molecules below are not the same, as they cannot be interconverted *via* rotation of single bonds.

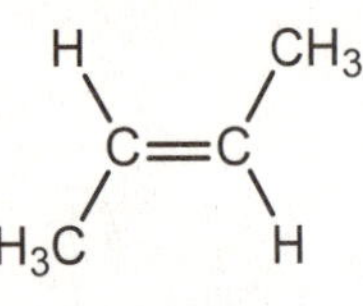

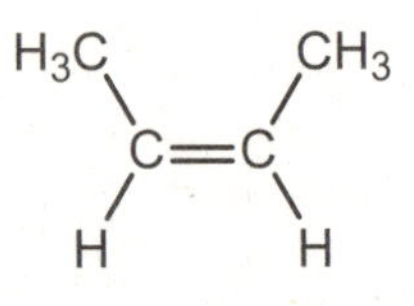

Critical Thinking Questions:

1. Recall the meanings of the terms *cis* and *trans* as applied to cycloalkanes. By analogy, label one of the molecules in Model 1 "*cis*-2-butene" and the other "*trans*-2-butene."
2. According to the definitions in the Information, what is the relationship between *cis*-2-butene and *trans*-2-butene?
3. Draw skeletal ("stick") representations of *cis* and *trans*-2-butene.
4. Draw skeletal ("stick") representations of *cis* and *trans*-3-hexene.

5. Identify the type of isomeric relationship between each pair of molecules below, from the following five choices (arranged from least to most similar): *not isomers, constitutional isomers, geometric isomers, conformers, identical.*

a.

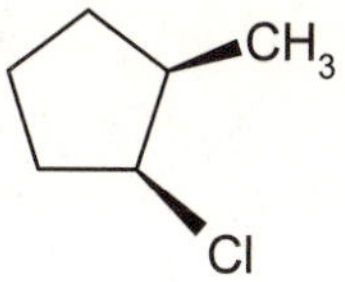

CH3 Cl and Cl H3C

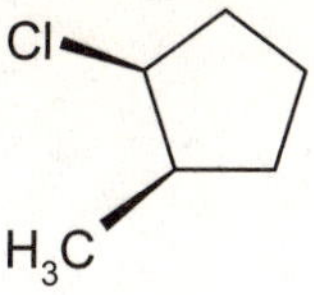

b.

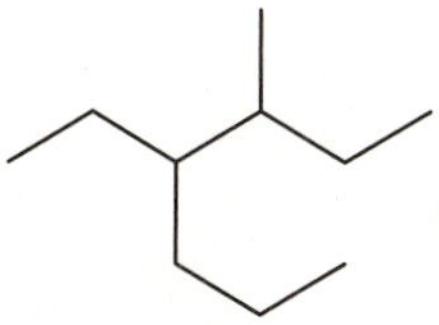

and

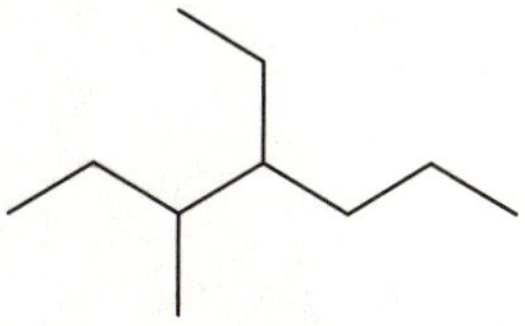

c.

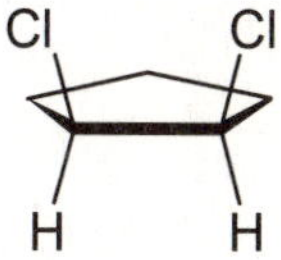

Cl Cl H H and H Cl Cl H

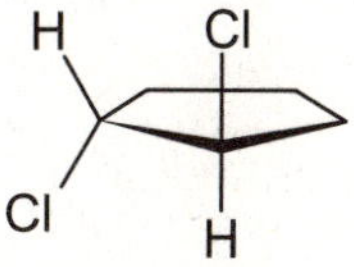

d.

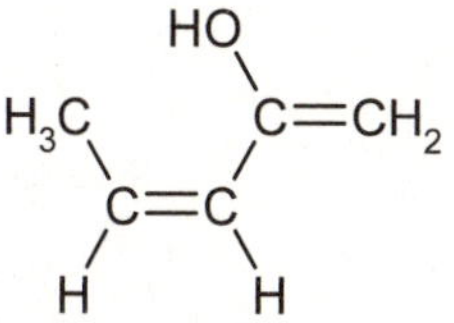

HO H3C C=CH2 C=C H H and H H C=C H3C C=CH2 HO

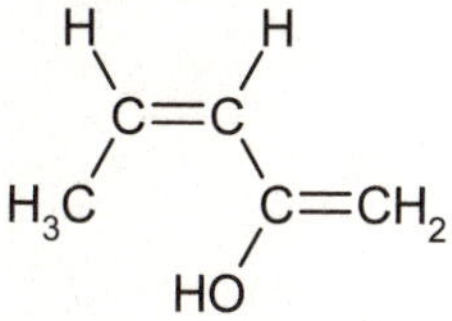

e.

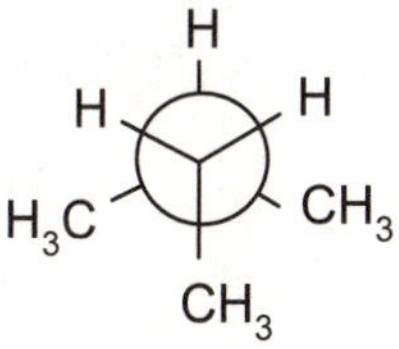

H H H H3C CH3 CH3 and CH3 H3C H H H CH3

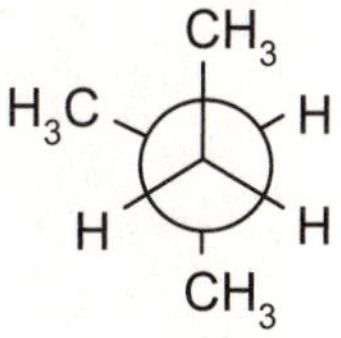

f.

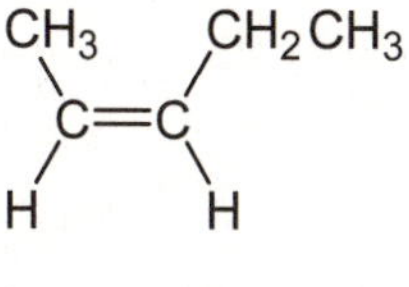

CH3 CH2CH3 C=C H H and

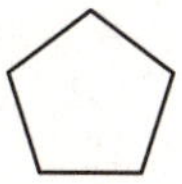

g.

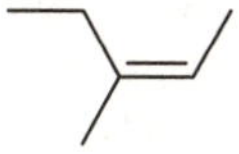

and

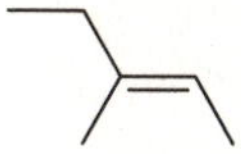

h.

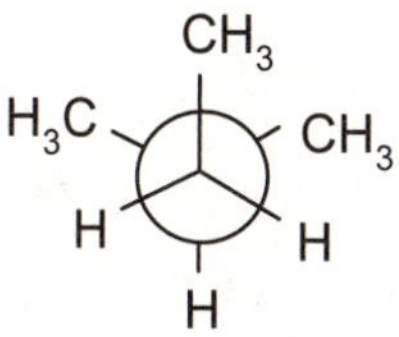

CH3 H3C CH3 H H H and H H H H CH3 CH3

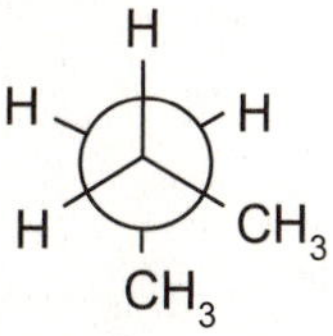

i.

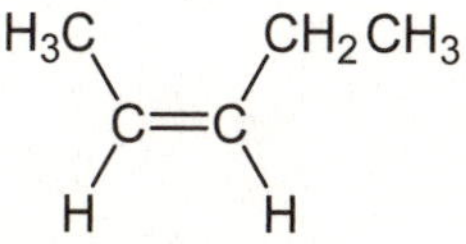

H3C CH2CH3 C=C H H and H CH3 C=C CH3CH2 H

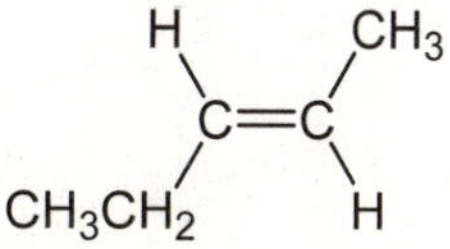

Exercises:

1. Redraw the structures from CTQ 5b as condensed structures.

2. Redraw the structures from CTQ 5e as condensed structures.

3. Redraw the structures from CTQ 5g as condensed structures.

4. Redraw the structures from CTQ 5h as condensed structures.

5. Redraw the structures from CTQ 5d as skeletal ("stick") structures.

6. Name the molecules in CTQ 5abceh.

   a.

   b.

   c.

   e.

   h.

7. Draw a representation (any style) of 1-butene. Can this molecule have *cis* and *trans* isomers? If not, explain. If so, draw them.

8. Read the assigned pages in your textbook and work the assigned problems.

Properties of Organic Molecules

(Which organic molecules are soluble in water?)

Model: Polarity and properties of organic molecules

Bond polarities are determined by the electronegativity difference between the bonded atoms. Recall that C-H bonds are nonpolar, but C-N and C-O bonds are polar. Since water is a polar solvent, the more polar an organic molecule, the more soluble it will be in water.

Table 1: Polarity, solubility and boiling points of selected compounds*

Alkane	Structure	Molar mass, g/mol	Dipole moment, Debyes	Boiling point, °C	Water solubility, grams per 100 mL H_2O
propane	$CH_3CH_2CH_3$	44	0	-42	0.007
butane	$CH_3CH_2CH_2CH_3$	58	0	0	0.006
pentane	$CH_3CH_2CH_2CH_2CH_3$	72	0	36	0.04
hexane	$CH_3(CH_2)_4CH_3$	86	0	69	0.001
heptane	$CH_3(CH_2)_5CH_3$	100	0	98	0.01

Alcohol	Structure	Molar mass, g/mol	Dipole moment, Debyes	Boiling point, °C	Water solubility, grams per 100 mL H_2O
ethanol	CH_3CH_2OH	46	1.7	78	∞
1-propanol	$CH_3CH_2CH_2OH$	60	1.7	82	∞
1-butanol	$CH_3CH_2CH_2CH_2OH$	74	1.67	118	6.3
1-pentanol	$CH_3(CH_2)_4OH$	88	1.7	137	2.7
1-hexanol	$CH_3(CH_2)_5OH$	102	1.8	157	0.6
1-heptanol	$CH_3(CH_2)_6OH$	116	1.7	176	0.1

Ether	Structure	Molar mass, g/mol	Dipole moment, Debyes	Boiling point, °C	Water solubility, grams per 100 mL H_2O
dimethyl ether	CH_3OCH_3	46	1.3	-23	∞
diethyl ether	$CH_3CH_2OCH_2CH_3$	74	1.15	35	6.9
dipropyl ether	$CH_3(CH_2)_2O(CH_2)_2CH_3$	102	1.2	89	0.25

Amine	Structure	Molar mass, g/mol	Dipole moment, Debyes	Boiling point, °C	Water solubility, grams per 100 mL H_2O
propyl amine	$CH_3CH_2CH_2NH_2$	59	1.3	48	∞
butyl amine	$CH_3CH_2CH_2CH_2NH_2$	73	1.3	77	∞
hexyl amine	$CH_3(CH_2)_5NH_2$	101	–	131	1.2
octyl amine	$CH_3(CH_2)_7NH_2$	129	–	180	0.02

* Sources: CRC Handbook of Chemistry and Physics, 47th ed., 1967; ChemFinder.com; IPCS INCHEM, www.inchem.org; Korea thermophysical properties Data Bank, www.cheric.org/kdb/ [accessed June 2006]

Notes: The dipole moment is a measure of the polarity of a molecule; ∞ means infinitely soluble (*i. e.,* the liquids are miscible); a dash means the data were unavailable.

Critical Thinking Questions:

1. Considering the dipole moments of the molecules in Table 1, which functional group is most **nonpolar**?

2. In general, do polar compounds have **higher** or **lower** boiling points than nonpolar compounds (circle one)?

3. For each functional group listed below, indicate how the boiling point changes as the molar mass increases.

   a. alkane

   b. alcohol

   c. amine

4. The boiling point of a liquid increases as the attractions between molecules increase. These attractions are called intermolecular forces. Based on your answers to CTQ 3, how do the intermolecular forces between molecules change as the molar mass increases?

5. Find one molecule from each functional group (alkane, alcohol, ether, amine) with roughly the same molar mass (within 5 g/mol), and write their names below. Rank these compounds from the highest to lowest boiling point.

6. Repeat CTQ 5 with another set of four compounds.

7. Based on relative boiling points, write the numbers from 1 to 4 under the functional group names below, with the number 1 being the group with the most intermolecular attractions, and 5 being the least.

   alkane   alcohol   ether   amine

8. For each functional group below, circle each type of bond contained in the molecule. You may refer to Table 1.

functional group	Type of bond			
alkane	C-H	C-O	O-H	N-H
alcohol	C-H	C-O	O-H	N-H
ether	C-H	C-O	O-H	N-H
amine	C-H	C-O	O-H	N-H

9. Based on your answer to CTQ 8, which two types of bonds are present in the molecules with the strongest intermolecular attractions?

Information:

The types of bonds you identified in CTQ 8 can exhibit what is known as **hydrogen bonding**. A "hydrogen bond" is simply a particularly strong attraction between the bonded hydrogen atom and a lone pair on another atom. This attraction causes molecules to stick together, but is much weaker than a covalent bond (up to one-tenth as strong), and so can be broken and reformed continually at room temperature.

Water has O-H bonds, and the O has two lone pairs, meaning that water meets the requirements for hydrogen bonding. Therefore, water is particularly suitable for dissolving organic molecules that can exhibit hydrogen bonding, since the water and organic molecule can "hydrogen-bond" together.

Consider the molecule ethanol, CH_3CH_2OH. It has one polar functional group (the –OH) and a two-carbon nonpolar alkyl group (CH_3CH_2-). Since the nonpolar group is not attracted to the water, it is reasonable to say that the –OH is the reason that ethanol is miscible with water.

Critical Thinking Question:

10. Suppose that we consider anything over about 1 gram per 100 mL water to be "soluble." Then, considering the water solubilities of the alcohols, ethers, and amines in Table 1, the presence of one polar functional group is sufficient to dissolve a molecule containing about how many nonpolar carbons? Circle one of the following choices:

1-2   3-4   5-6   7-8

Table 2: Water solubility of selected ketones

Ketone	Structure	Water solubility, grams per 100 mL H_2O	Ketone	Structure	Water solubility, grams per 100 mL H_2O
acetone	$H_3C-C(=O)-CH_3$	∞	2-hexanone	$H_3C-C(=O)-(CH_2)_3CH_3$	1.4
2-butanone	$H_3C-C(=O)-CH_2CH_3$	25.6	2-heptanone	$H_3C-C(=O)-(CH_2)_4CH_3$	0.4
2-pentanone	$H_3C-C(=O)-CH_2CH_2CH_3$	4.3			

Exercises:

1. Consider the water solubilities of the ketones shown in Table 2. Are they consistent with your answer to CTQ 10? Write a sentence that generalizes how many nonpolar carbons can be dissolved by one polar functional group.

2. A hydrogen bond is usually depicted by a dashed or dotted line. Circle the picture that correctly represents a hydrogen bonding interaction.

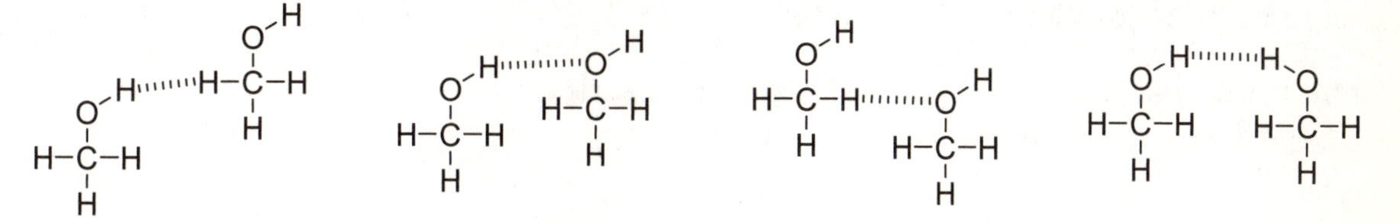

3. Describe what is wrong with each of the other three pictures in Exercise 1.

4. Draw a representation of one water molecule participating in a hydrogen bond with another water molecule. Then place the symbols δ+ and δ− near each atom to indicate the polarity of the bonds.

5. Without looking up any information in a table, identify the molecule in each group that would have the highest boiling point, and explain your answer.

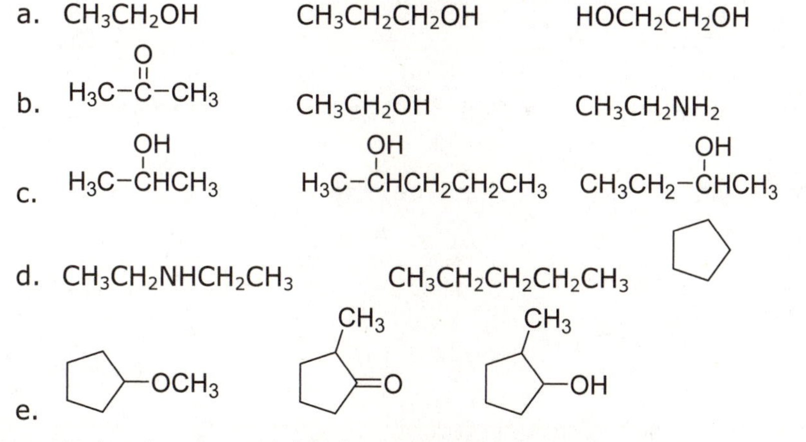

6. Benzoate ion is very soluble in water, but benzoic acid is not. Based on this information which species do you think has a larger dipole moment—benzoic acid or benzoate? Explain.

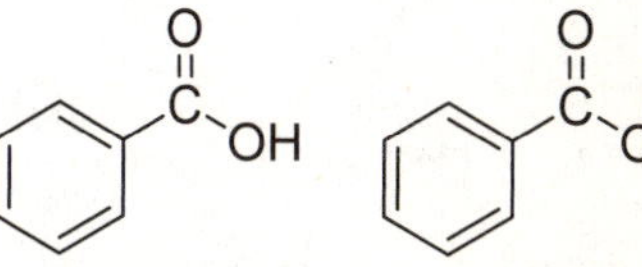

benzoic acid benzoate

7. Read the assigned pages in your text, and work the assigned problems.

Reactions of Organic Molecules

(What are the main types of organic reactions in biological systems?)

Information: Organic reactions common in biochemistry

There are seven common reaction classes of organic molecules that we will consider. The first class—acid-base reactions—is one we have seen before. Then there are three pairs of reactions that are opposites of each other—addition and elimination; reduction and oxidation; condensation and hydrolysis.

Table 1: Seven common types of organic reactions

1. **acid-base**: transfer of a proton (H^+)

   - example:

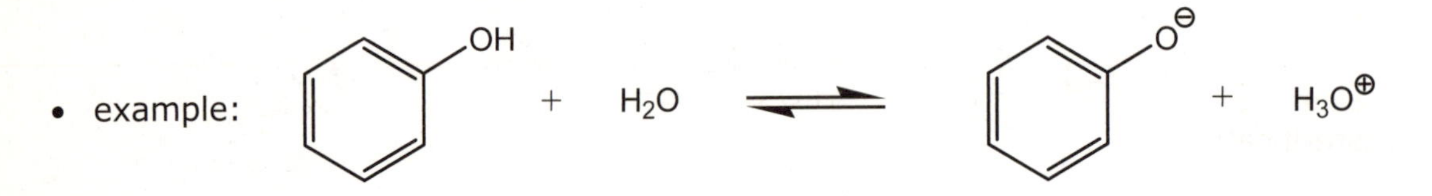

2. **addition**: addition of a small molecule (usually H_2O) across a double bond

   - example (hydration): $CH_3{-}CH{=}CH_2 + H_2O \xrightarrow{H^+ \text{ catalyst}} CH_3{-}CH(OH){-}CH_2(H)$

3. **elimination**: removal of a small molecule (often H_2O) to form a double bond

   - example (dehydration): $CH_3{-}CH(OH){-}CH_2(H) \xrightarrow[\text{and heat}]{\text{conc. acid}} CH_3{-}CH{=}CH_2 + H_2O$

4. **reduction**: addition of 2 H atoms to or removal of an O atom from a molecule. Can be symbolized "[r]"

   - example: $CH_3{-}CH{=}CH_2 + H_2 \xrightarrow{\text{Pt catalyst}} CH_3{-}CH_2{-}CH_3$

5. **oxidation**: addition of an O atom to or removal of 2 H atoms from a molecule. Can be symbolized "[o]"

   - example with primary alcohol: forms an aldehyde, **then** an acid

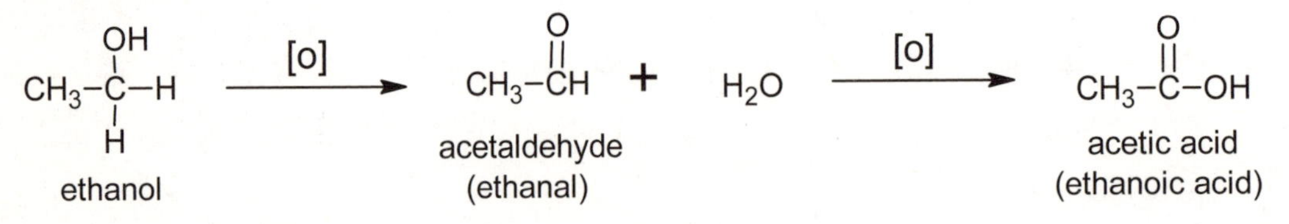

   - example with secondary alcohol: forms an ketone

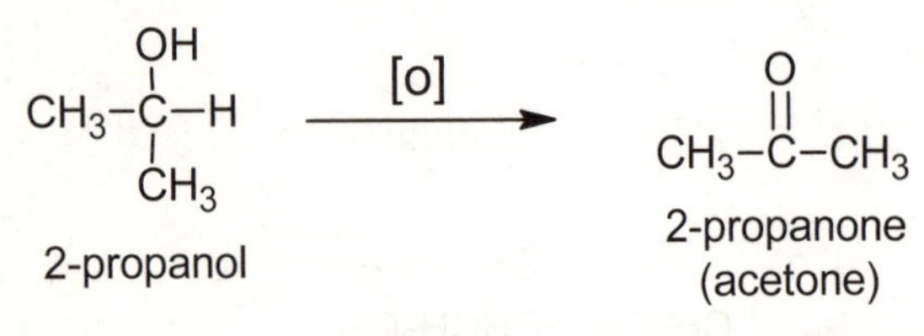

- example with tertiary alcohol: no reaction

$$CH_3-\underset{CH_3}{\overset{OH}{\underset{|}{\overset{|}{C}}}}-CH_3 \xrightarrow{[o]} \text{NR (no H's on alcohol carbon to be removed)}$$

tert-butyl alcohol

- example with thiol: forms a disulfide

$$2\,CH_3\text{-}SH \xrightarrow{[o]} CH_3\text{-}S\text{-}S\text{-}CH_3\ (+\ H_2O)$$

methanethiol → dimethyl disulfide

6. **condensation**: coupling of two molecules with the loss of a small molecule (usually H_2O)

- example (esterification):

$$CH_3-\overset{O}{\overset{||}{C}}\text{-}\boxed{OH\ +\ H}\text{-}O\text{-}CH_3 \xrightleftharpoons{H^+\ \text{catalyst}} CH_3-\overset{O}{\overset{||}{C}}-OCH_3\ +\ H_2O$$

acetic acid (an acid) + methanol (an alcohol) → methyl acetate (an ester)

7. **hydrolysis**: splitting a molecule in two with the addition of water

- example (ester hydrolysis):

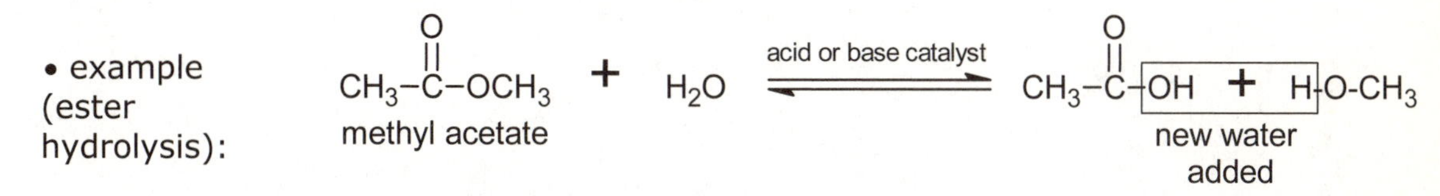

Just as it is beneficial to make a summary of the types of reactions, it is also useful to make a list of the functional groups that commonly react by each of the seven pathways. Such a summary is shown in Table 2.

Table 2: Functional groups that commonly react by the seven reaction pathways

acid-base	– carboxylic acids, phenols (acids)
	– amines (bases)
addition	– alkenes, alkynes
elimination	– alcohols
reduction	– alkenes, ketones
addition	– carboxylic acids, phenols (acids)
oxidation	– aldehydes, primary and secondary alcohols
condensation	– acid + alcohol, acid + amine
hydrolysis	– esters, amides (peptides)

Critical Thinking Questions:

1. For the acid-base reaction in Table 1, identify the acid and base on the reactant side and the conjugate acid and conjugate base on the product side. Draw a line to connect conjugate acid-base pairs together.

2. Tell which of the seven common types of organic reactions is illustrated by each of the following.

a. $CH_3CH_2-\overset{O}{\overset{||}{C}}-CH_2CH_3\ +\ H_2 \xrightarrow{Pt} CH_3CH_2-\overset{OH}{\overset{|}{C}}HCH_2CH_3$

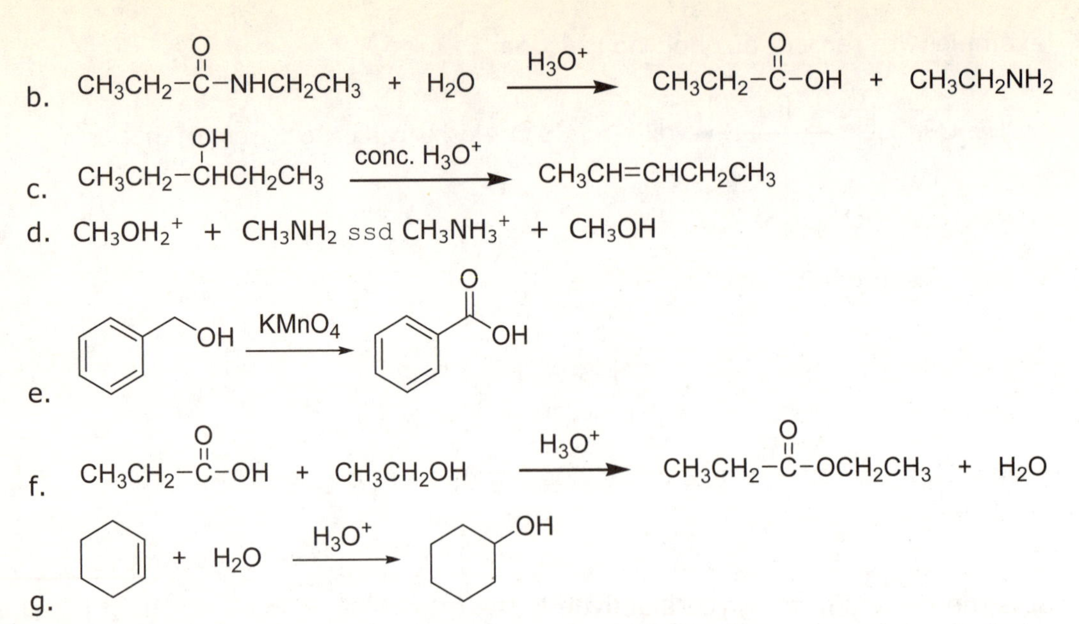

3. Classify each of the molecules below as a primary, secondary, or tertiary alcohol.

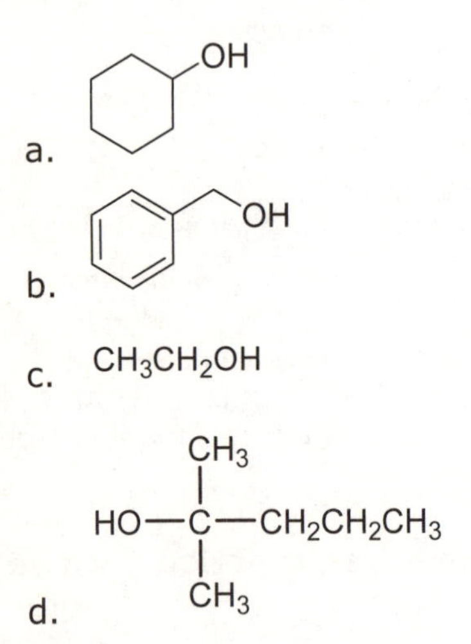

4. For each of the alcohols in Exercise 3, show the oxidation products, if any. If the oxidation takes place in two steps, show the product of each step.

Exercises:

1. Each of the following carboxylic acids can donate a hydrogen atom to water. Complete a chemical reaction for each.

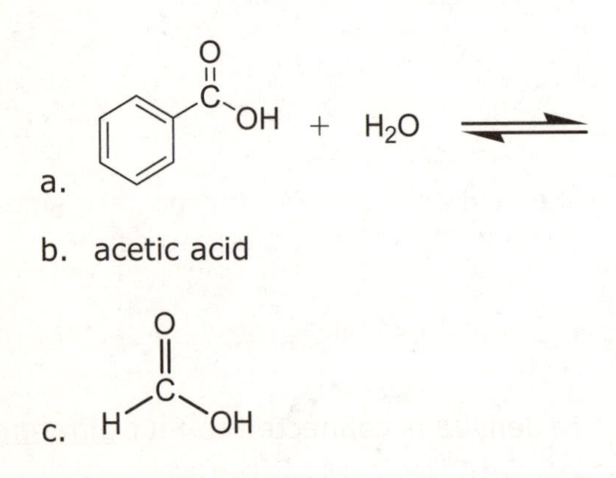

2. Read the assigned pages in your textbook and work the assigned problems.

Carbohydrates

(What makes a sugar?)

Model 1: Glyceraldehyde

Simple sugars have a molecular formula of $C_n(H_2O)_n$, meaning that for *n* carbon atoms in the sugar, there are *n* oxygen atoms and *2n* hydrogen atoms. The simplest sugar (a monosaccharide) is glyceraldehyde, with the molecular formula $C_3H_6O_3$ and the structure shown below.

```
      O    OH
      ||   |
H—C—CH-CH2OH
```

Critical Thinking Questions:

1. Circle the aldehyde functional group in the structure of glyceraldehyde above.
2. What is the value of *n* for glyceraldehyde in the formula $C_n(H_2O)_n$?
3. Hypothesize on the origin of the term *carbohydrate*.

4. Using a molecular model kit, make a molecule of glyceraldehyde. (In many model kits, black = C, red = O, white = H. Use short bonds for single bonds and <u>two</u> of the longer, flexible bonds for double bonds.)

   Hold your model with the aldehyde carbon at the top and the CH_2OH at the bottom, and arrange it so that the middle carbon is closer to you than the other two carbons.

   Now, compare your model with the two structures below. Circle the one that matches your model. (Remember what the dashes and wedges mean?)

5. Now make <u>a second</u> model of glyceraldehyde, so that you have one of each of the two structures shown above. Can you rotate or twist them so that all the atoms are in the same place in both molecules? ______

Model 2: Types of stereoisomers

Stereoisomers are molecules with the same connectivity, but different arrangements in space. This category includes geometric (*cis-trans*) isomers and a type of isomers we have not seen before called *enantiomers*. Enantiomers are a pair of non-identical molecules that are mirror images of each other.

Whenever a carbon is bonded to **four dissimilar groups**, it is said to be *chiral*. An object or molecule that is chiral is not identical to its mirror image. The term chiral literally means "handedness."

Critical Thinking Question:

6. Confirm that the center carbon of either glyceraldehyde is connected to four <u>different</u> (or dissimilar) groups.

7. Are the two forms of glyceraldehyde **geometric isomers** or **enantiomers** (circle one)? Explain in one complete sentence.

Figure 1: Memory device for D-glyceraldehyde

To distinguish the two isomers of glyceraldehyde in CTQ 4, chemists call the one on the left L-glyceraldehyde and the one on the right D-glyceraldehyde. One way to remember which is which is to imagine a line from C1 to C3 through the middle OH. The D isomer makes the letter "D." Nearly all common sugars of biological significance have the D configuration.

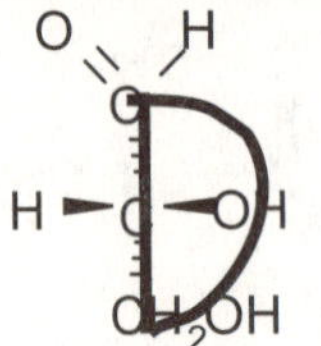

Figure 2: Wedge-and-dash structures and Fisher projections of glyceraldehyde

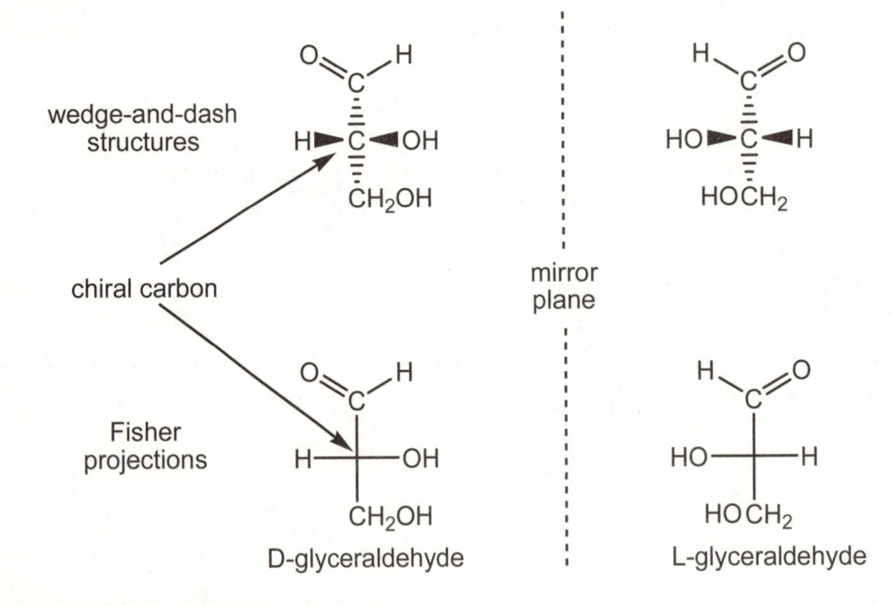

Critical Thinking Question:

8. The Fisher projection of L-glyceraldehyde shown below does not exactly match the Fisher projection in Figure 2. Explain how the two structures are different, and why both are correctly named. You may wish to look at and manipulate your model.

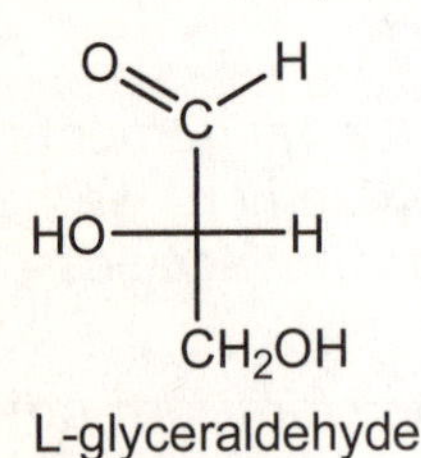

Information:

The center carbon of glyceraldehyde is connected to four different (or dissimilar) groups. This makes it a **chiral** carbon and leads to the two enantiomers. Each additional chiral carbon doubles the number of pairs of enantiomers, so that in 6-carbon carbohydrates with four chiral carbons, there are 16 possible isomers ($2^4 = 16$). In the Fisher projection, the last (bottom) chiral carbon determines whether the isomer is L or D.

Definitions:

aldose – a simple sugar containing an aldehyde
ketose – a simple sugar containing a ketone

Most of the common sugars have 5 or 6 carbons, and so are called **pentoses** or **hexoses**. You can mix the terms with **aldose** and **ketose** to get names such as **ketopentose** and **aldohexose**.

Figure 3: Common simple sugars

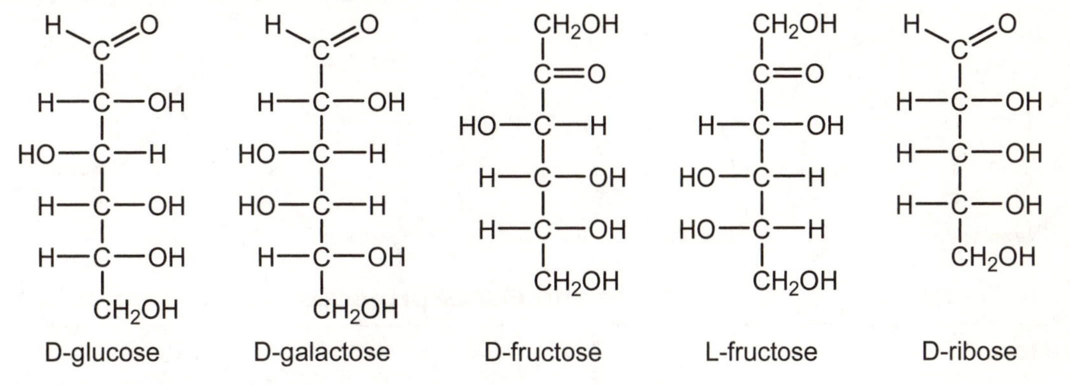

Critical Thinking Questions:

9. Label each sugar in Figure 3 with the appropriate name from this list: aldohexose, aldopentose, ketohexose, ketopentose.
10. Circle *each* chiral carbon in Figure 3. (Hint: There are 17 total).
11. Consider the two isomers of fructose in Figure 3. What is the difference between the L and D isomer of a simple sugar?

Information:

The simple 5- and 6-carbon sugars often react with themselves to make cyclic structures. When this happens, a new functional group is formed—a *hemiacetal*. A hemiacetal has an alcohol (-OH) and an ether (-OR) attached to the same carbon. An example is shown in Figure 4.

Figure 4: Cyclization of D-glucose to make α-D-glucose.

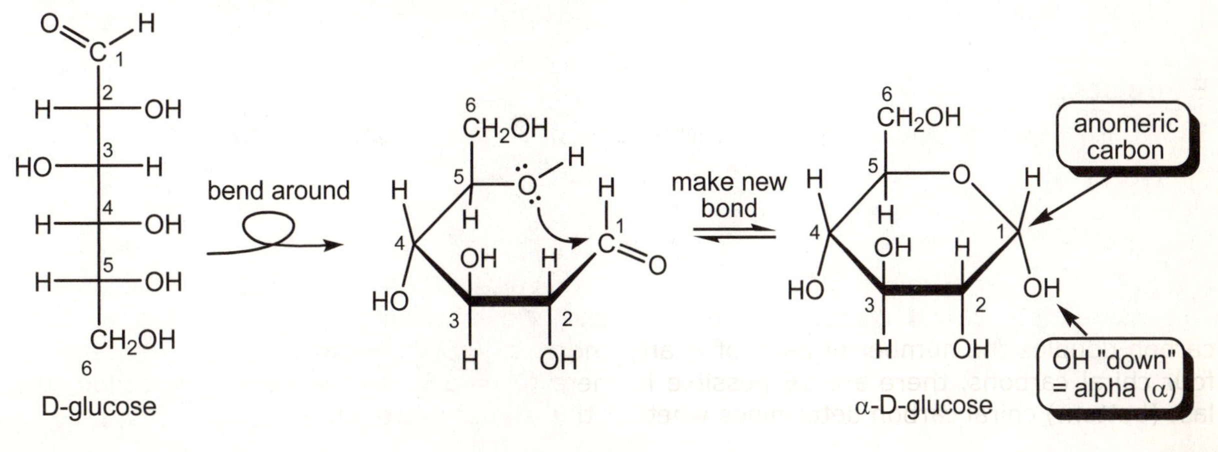

When this happens, a new chiral carbon is created. This carbon (the **anomeric** carbon) is designated α or β depending on whether its OH is "down" or "up" when the cyclic structure is drawn in the orientation shown in Figure 4. In disaccharides and polysaccharides, simple sugars are connected together with at least one of them bonding at the anomeric carbon. Bonds to the anomeric carbon are called **glycosidic bonds**.

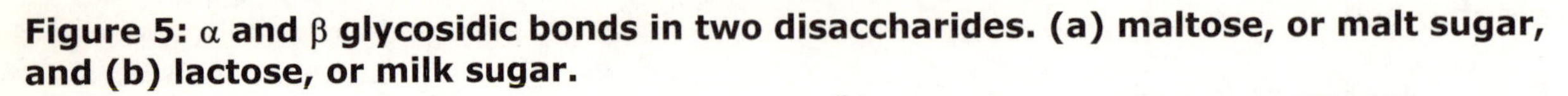

Figure 5: α and β glycosidic bonds in two disaccharides. (a) maltose, or malt sugar, and (b) lactose, or milk sugar.

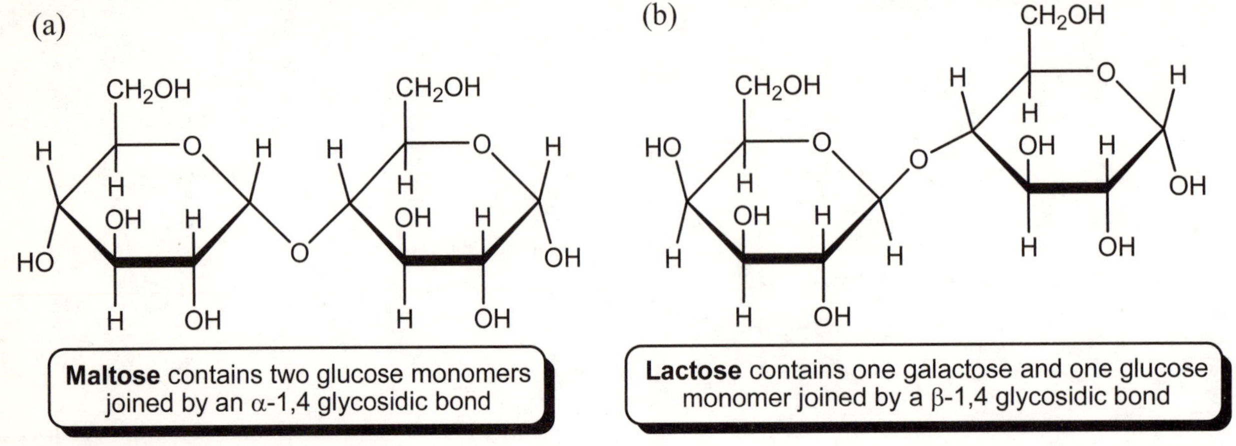

Critical Thinking Questions:

12. Consider Figures 4 and 5. Explain what is meant by "α," "1," and "4" in the term α-1,4 glycosidic bond.

13. Explain what is meant by "β," "1," and "4" in the term β-1,4 glycosidic bond.

14. By extension, explain what would be meant by the term α-1,6 glycosidic bond.

Exercises:

1. What is the difference between the structures of D-glucose and D-galactose?

2. Considering your answer to Exercise 1 and Figures 4 and 5, explain how you can determine which part of the disaccharide lactose in Figure 5 came from the glucose monomer, and which came from galactose. Then label the Figure accordingly.

3. Find descriptions in your textbook of the structures of the polysaccharides amylose, amylopectin, and cellulose. Describe the similarities and differences among these polymers. Be sure to consider the types of glycosidic bonds as examined in CTQs 12-14.

4. Read the assigned pages in your textbook and work the assigned problems.

Lipids

(What are the components of cell membranes?)

Information: Fatty Acids

Fatty acids have a carboxyl group "head" and a hydrocarbon "tail" that is 11-19 carbons in length.

Figure 1: Structure of palmitoleic acid and its conjugate base

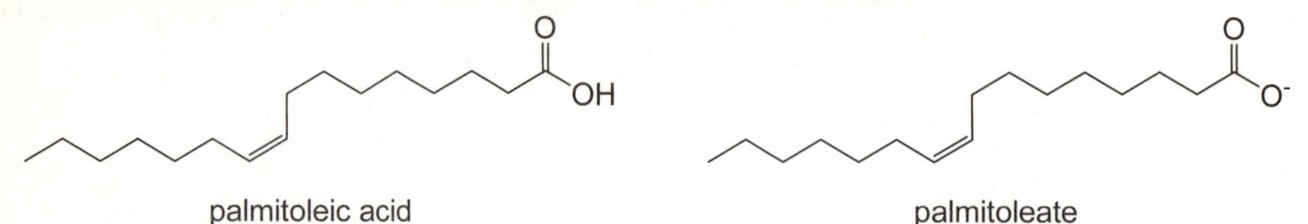

Critical Thinking Questions:

1. Which would be more water soluble:—palmitoleic acid, or sodium palmitoleate? Explain.

2. Select a carbon in Figure 1 that is in a C=C bond, and circle it. To how many other **atoms** is this carbon covalently bonded?

Information:

We say that a carbon is **saturated** when it is bonded to 4 different atoms. A C=C in a molecule is referred to as a **site of unsaturation**. Fatty acids are often **unsaturated**, that is, they contain one or more *cis* carbon-carbon double bonds.

Critical Thinking Questions:

3. Compare two or more fatty acids from Table 1 that have the same number of carbons (and roughly the same molar mass).
   a. How does the melting point change as the number of sites of unsaturation increases?

   b. How do the intermolecular attractions between molecules change as the number of sites of unsaturation increases?

   c. Given the shapes of the molecular structures shown in Table 1, devise an explanation for the effect that you described in CTQ 3b.

4. Give a name of a fatty acid from Table 1 that fits the class listed.
   a. a saturated fatty acid
   b. a monounsaturated fatty acid
   c. a polyunsaturated fatty acid

5. Which class of fatty acid listed in CTQ 4 would be a solid at room temperature?

Information: Lipids

Table 1: Structures and melting points of some common fatty acids*

Alkane	Number of carbons	Structure	Molar mass, g/mol	Melting point, °C
palmitic acid	16	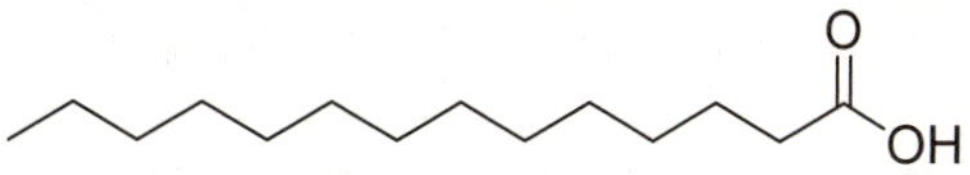 OH, O	256	62
stearic acid	18	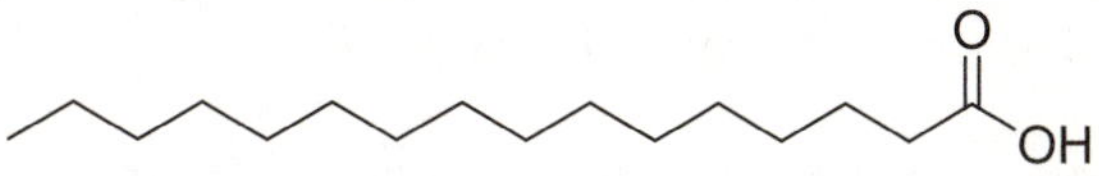 OH, O	284	69
arachidic acid	20	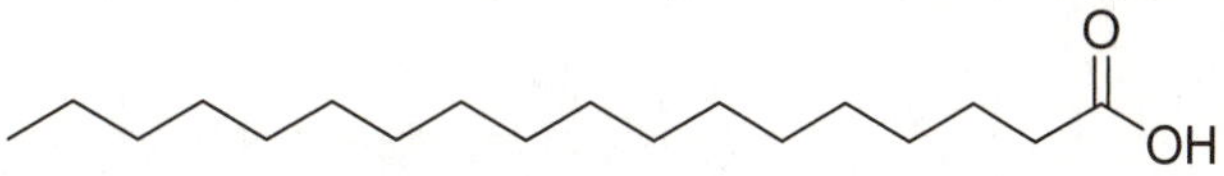 OH, O	313	75
palmitoleic acid	16	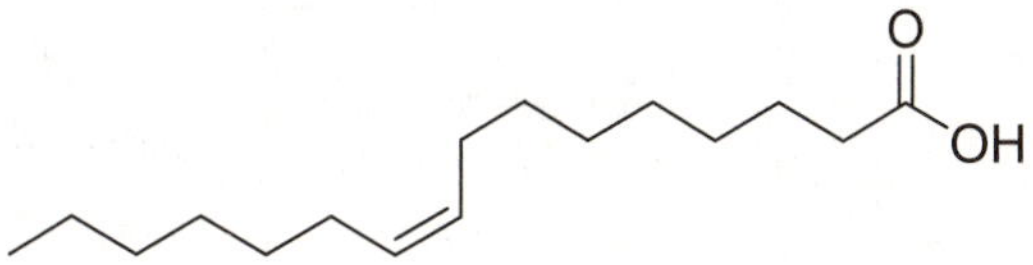 OH, O	254	0
oleic acid	18	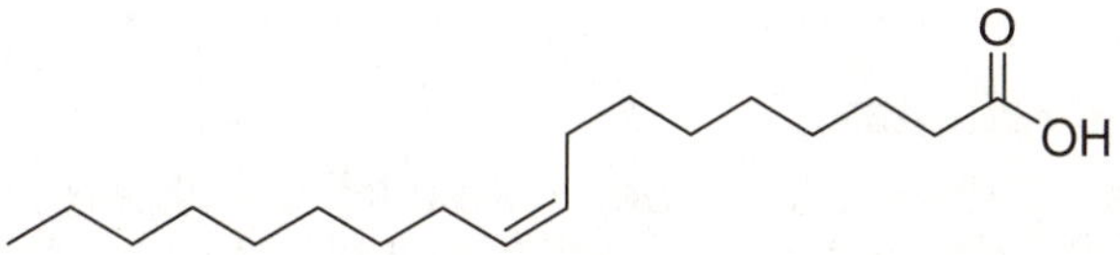 OH, O	282	13
linoleic acid	18	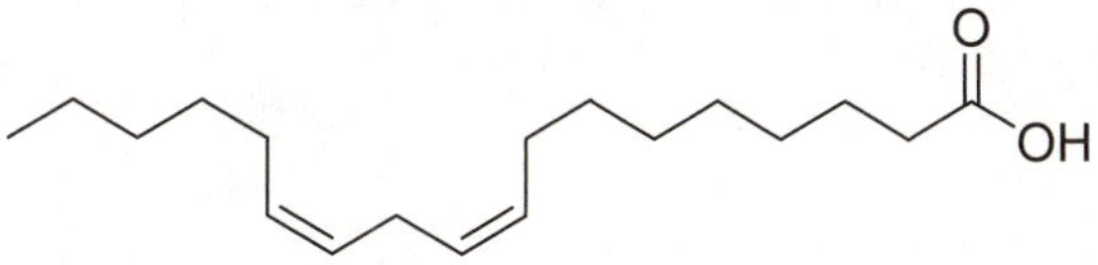 OH, O	280	-5
linolenic acid	18	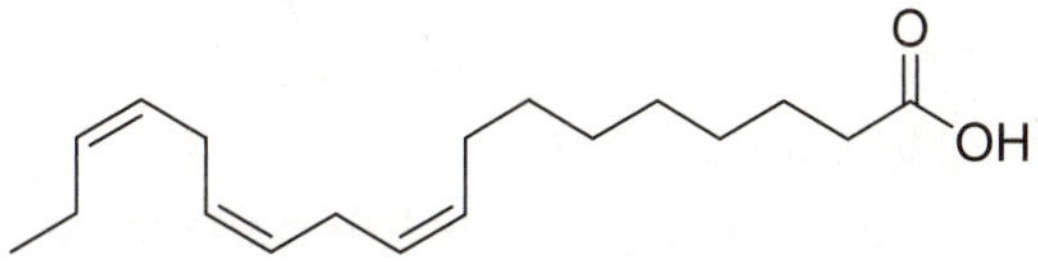 OH, O	278	-11
arachidonic acid	20	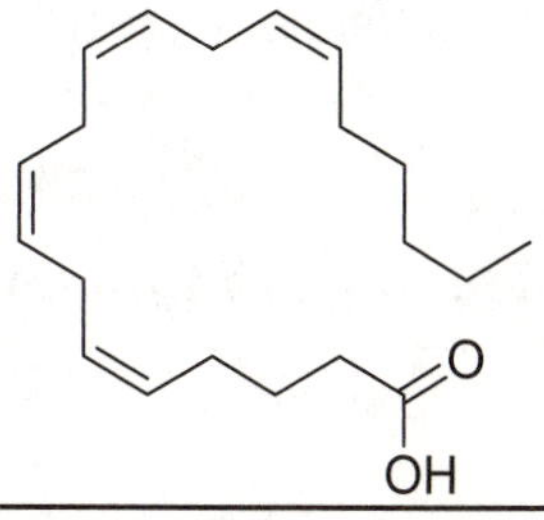 O, OH	304	-49

Prostaglandins are formed from the 20-carbon fatty acid arachidonic acid. Various prostaglandins act like hormones in the body. Steroids and pharmaceuticals such as nonsteroidal anti-inflammatory drugs (NSAIDs) inhibit the formation of prostaglandins responsible for producing inflammation and pain.

Waxes are modified fatty acids in which the hydrogen of the carboxyl head is replaced with a second hydrocarbon tail. Many plants and animals produce or secrete waxes as a protective, water-repellent barrier.

* Source: ChemFinder.com [accessed June 2006]

Figure 2: Carnauba wax can be isolated from palm trees.

Critical Thinking Question:

6. What organic functional group (other than alkane) is contained in a wax molecule?

Information:

Chemically, **fats and oils** are **triacylglycerols** (triglycerides). They are composed of a glycerol "backbone" esterified with three fatty acids.

Figure 2: Composition of a generic fat molecule (triacylglygerol)

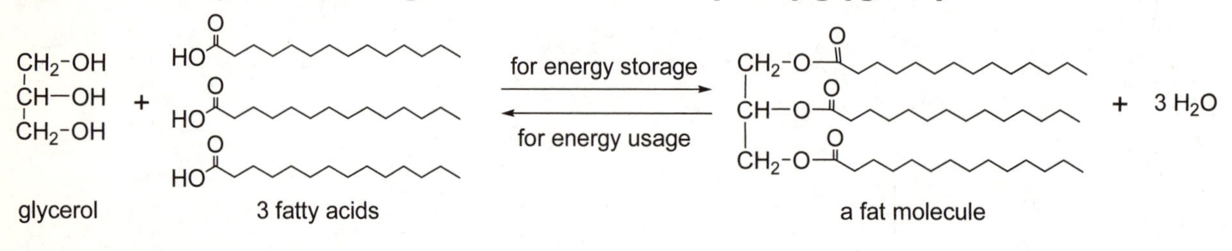

Critical Thinking Questions:

7. The main difference between a fat and an oil is that oils are liquids at room temperature. What conclusion could you make about the likely class of the three fatty acyl groups in an oil versus those in a fat?

Information: Cell Membranes

The main lipid component of membranes is **glycerophospholipids**. The structure of a glycerophospholipid is like a fat molecule (triacylglycerol), but with one fatty-acyl "tail" replaced with a phosphate group plus an amino alcohol. This gives the glycerophospholipid one very polar "head" group with **two** nonpolar "tail" groups.

Figure 4: A phosphatidylethanolamine, a typical glycerophospholipid

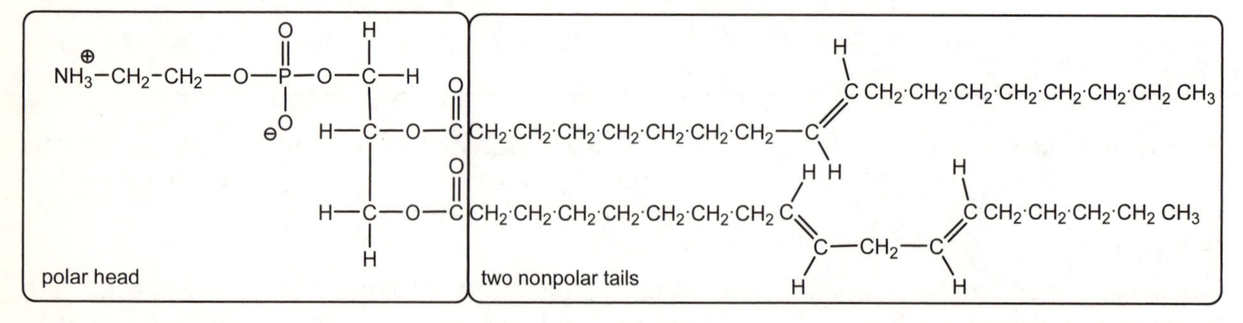

The other lipid component of membranes is the steroid **cholesterol**, shown in Figure 5. Other steroids with the same ring structure are hormones such as estrogen and testosterone.

Figure 4: Structure of cholesterol as commonly drawn (a), and in the typical "rigid" chair conformation (b)

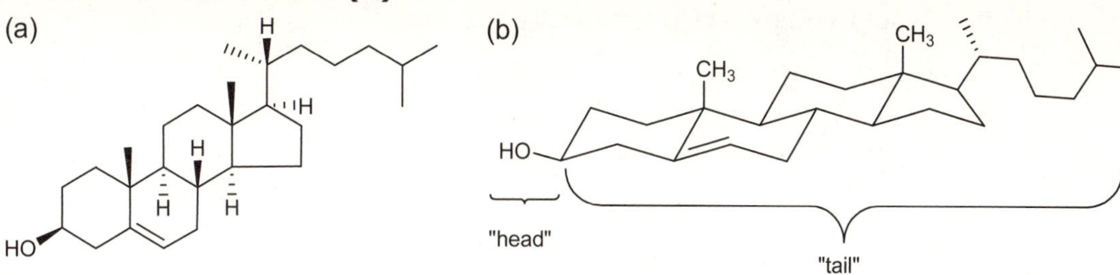

Cell membranes are composed of **lipid bilayers** with associated proteins. The bilayers are formed when lipids associates in two layers with the hydrophobic (nonpolar) tails on the inside, and the hydrophilic (polar) heads on the outsides. See Figure 3 for a schematic of a cell (plasma) membrane. (Notice in the cartoon in Figure 3 how each head has two tails?)

Cholesterol has a very small polar head with a large, nonpolar tail. Because of the four fused rings in the structure, cholesterol is very rigid, and therefore adds rigidity to the membrane.

Figure 3: Model of a lipid bilayer as a cell membrane. Membrane proteins float in a "sea" of lipids, but cannot undergo transverse movement ("flip-flop").

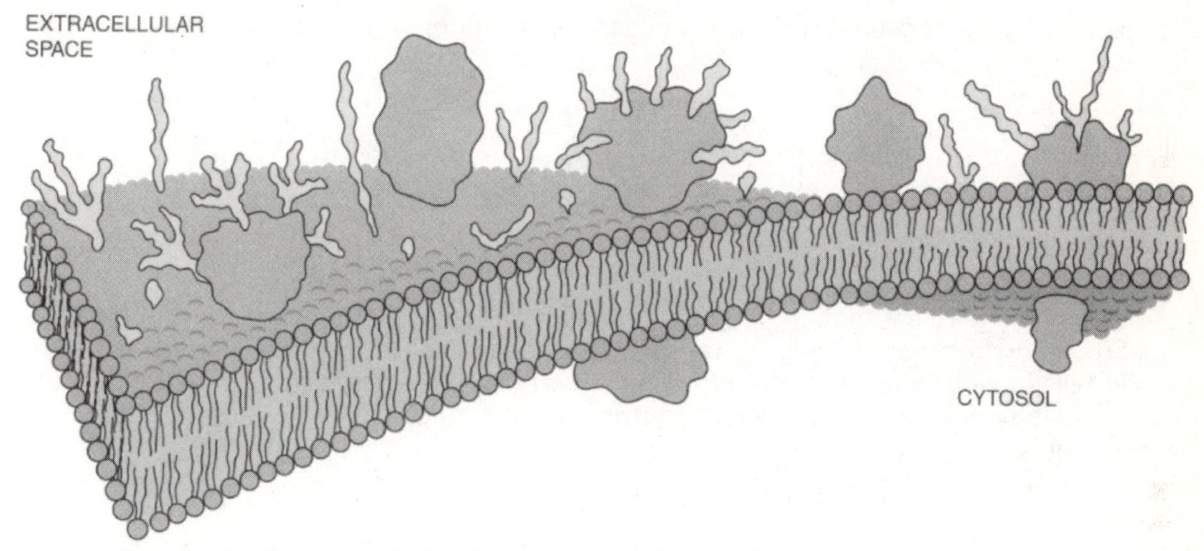

Critical Thinking Questions:

8. A cell membrane can vary in terms of how "fluid" or flexible it is. Describe if you would expect each of the following changes to make a particular membrane more fluid or more rigid. Explain your choices.

   a. an increase in the cholesterol content

   b. an increase in the fraction of unsaturated fatty-acyl tails in the membrane glycerophospholipids

9. Would you expect a reindeer living in northern Canada to have a cell membrane composition similar to a red-tailed deer in the southern United States? If so, explain why. If not, how would you expect the membranes to differ?

Exercises:

1. Describe the differences between triacylglycerols and glycerophospholipids.

2. Explain why a leaf coated with wax makes the leaf water-repellent.

3. A common ingredient in many prepared foods is *partially hydrogenated soybean oil.* "Partially hydrogenated" means that some (but not all) of the double bonds in the fatty acid tails have been converted to single bonds. In this process, some of the remaining *cis* double bonds are also converted to *trans* double bonds.
   a. What would be the purpose of converting some of the double bonds in the fatty acid tails to single bonds? (How would this affect the properties of the oil?)

   b. How can you reduce the amount of *trans* fats in your diet?

4. Read the assigned pages in the text and work the assigned problems.

Amino Acids and Proteins

(What does it mean for a protein to be denatured?)

Information:

The building blocks of proteins are α-**amino acids**, small molecules that contain a carboxylic acid and an amino group. The amino group is connected to the carbon next to the carboxyl group, designated the α carbon. There are 20 different amino acids found in proteins, differing only in the **side chain**.

Proteins are made of long chains of amino acids bonded together and folded into a particular shape. Proteins may be either fibrous or globular, and the specific shape of each protein is individualized to help it perform a specific function.

Table 1: Proteins perform many important functions

Function	Type	Examples
structure	fibrous	*microfilaments* are part of the cytoskeleton
catalysis	globular	*sucrase* is the enzyme that aids the hydrolysis of sucrose to fructose and glucose
contraction	globular and fibrous	*actin* and *myosin* in muscle fibers
transport	globular	*hemoglobin* in blood; various membrane proteins perform active transport

Model 1: General structure of an α-amino acid in acidic, neutral, or basic solution.

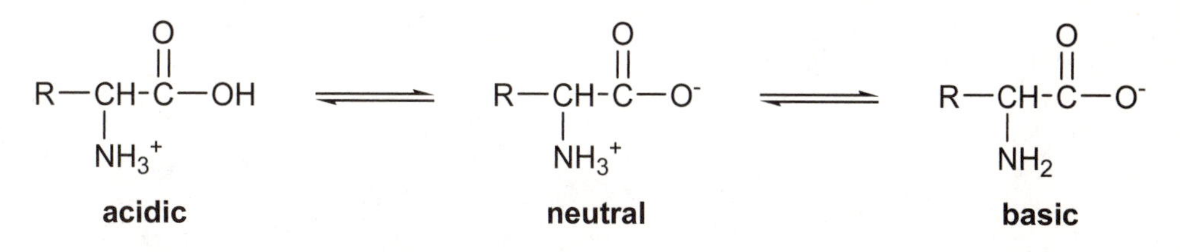

Amino acids contain both a carboxylic acid (proton donor) and a basic amino group (proton acceptor). The center "zwitterionic" form is commonly found in water solutions with a pH near neutrality. The side chain "R" may be one of 20 choices.

Critical Thinking Questions:

1. Using Table 2 on the following page to help you, draw the structure of the given amino acid as it would commonly exist under each of the following conditions.

   a. Valine; in the stomach at pH 1.5

   b. Serine; in the small intestine at pH 8

   c. Glutamate; in the blood plasma at pH 7.4

Table 2: Structures and abbreviations of the 20 standard amino acids in zwitterionic form, classified according to their side chain (R group) chemistry

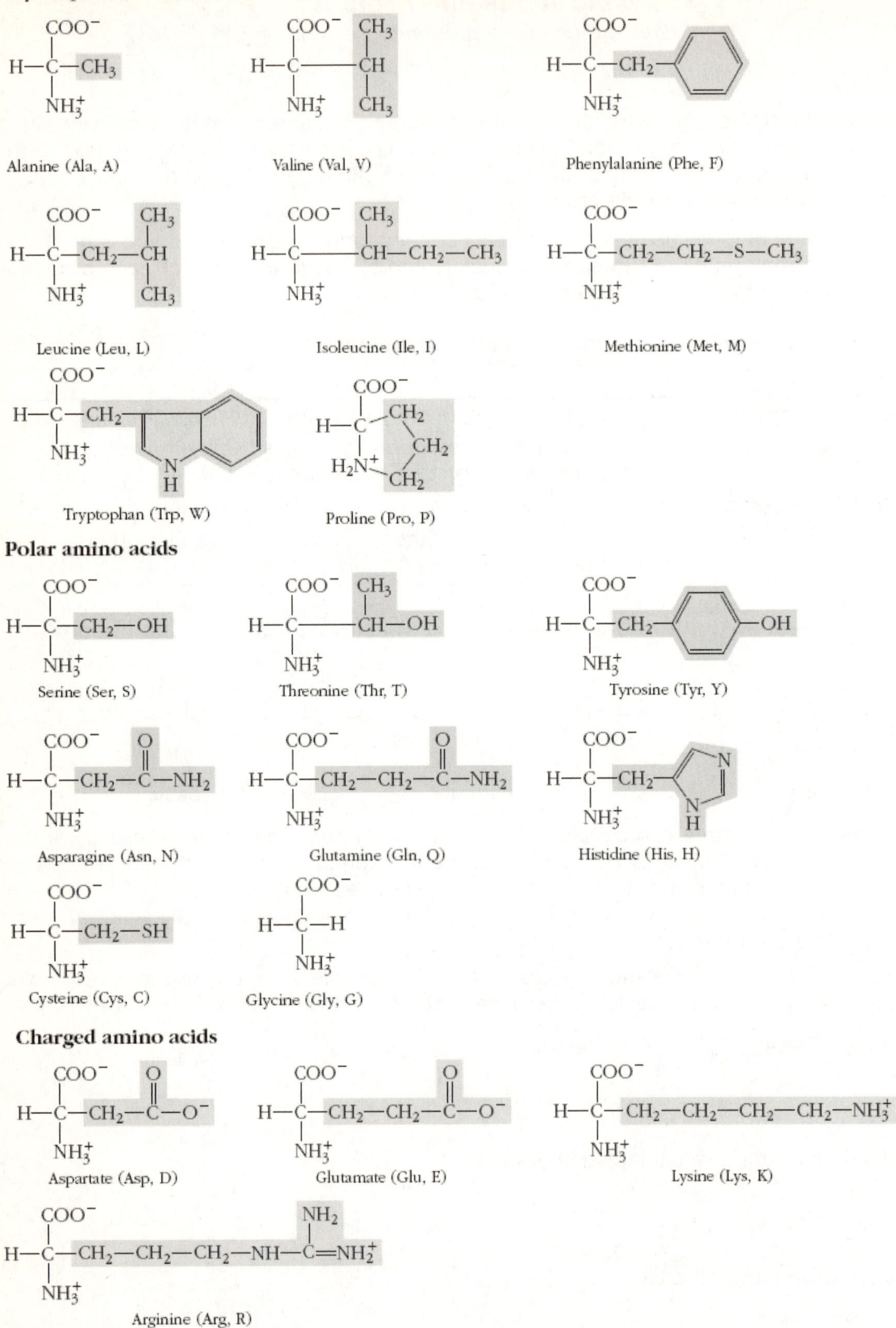

Critical Thinking Questions:

2. Which of the three classes of amino acids (hydrophobic, polar, charged) have R groups that would be attracted to water?

3. What is the opposite of hydrophobic?

4. Consider your answer to CTQ 3. To which of the three classes of amino acids could you apply this term?

Information:

The amino group of one amino acid and the acid group of another can undergo a condensation reaction to form a new bond, called either a **peptide bond** or an **amide bond**. The new molecule is called a dipeptide. More condensations can lead to tripeptides, tetrapeptides, *etc.*, and finally **polypeptides**.

Model 3: The condensation reaction of glycine and alanine to make the dipeptide glycylalanine.

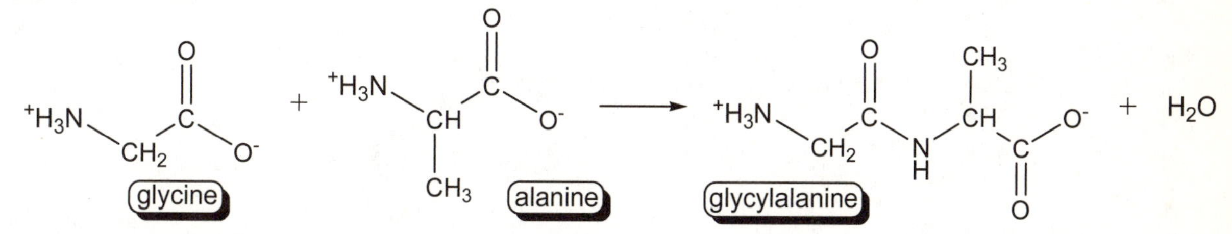

Critical Thinking Questions:

5. Draw a box around the part of the structure for glycylalanine in Model 3 that originated in **alanine**.

6. Draw another box around the part of the structure for glycylalanine in Model 3 that originated in **glycine**.

7. Draw an arrow to the bond that was formed in the condensation reaction to connect glycine to alanine. This bond is called a **peptide bond**. Label it in the model.

8. Would the dipeptide Ala-Gly (alanylglycine) be the same molecule as Gly-Ala (glycylalanine)? Explain.

Model 4: A tripeptide

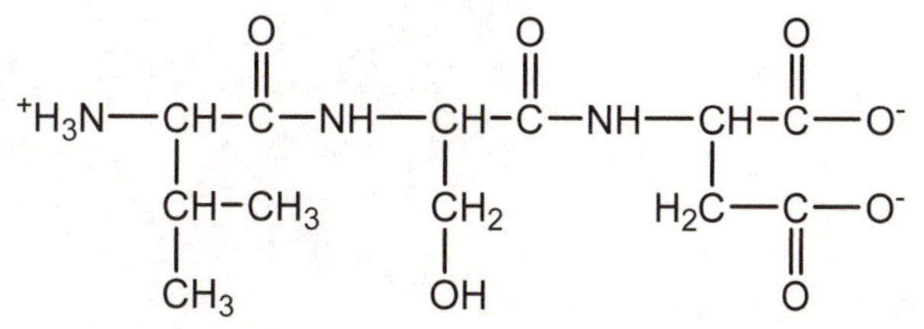

Critical Thinking Questions:

9. Circle the three side chains (R groups) in Model 4. What three amino acids condensed together to make the tripeptide? Label them in the Model.

10. Construct an explanation for why the central amino acid in Model 4 is called an amino acid *residue*.

Information:

A polypeptide with a specific biological function is called a **protein**. The amino acid sequence of a protein is called the **primary structure**. Protein structure is usually classified into four levels.

Figure 1: Levels of protein structure in hemoglobin.

Primary structure
The sequence of amino acid residues

Secondary structure
The localized conformation of the polypeptide backbone

Tertiary structure
The three-dimensional structure of an entire polypeptide, including all its side chains

Quaternary structure
The spatial arrangement of polypeptide chains in a protein with multiple subunits

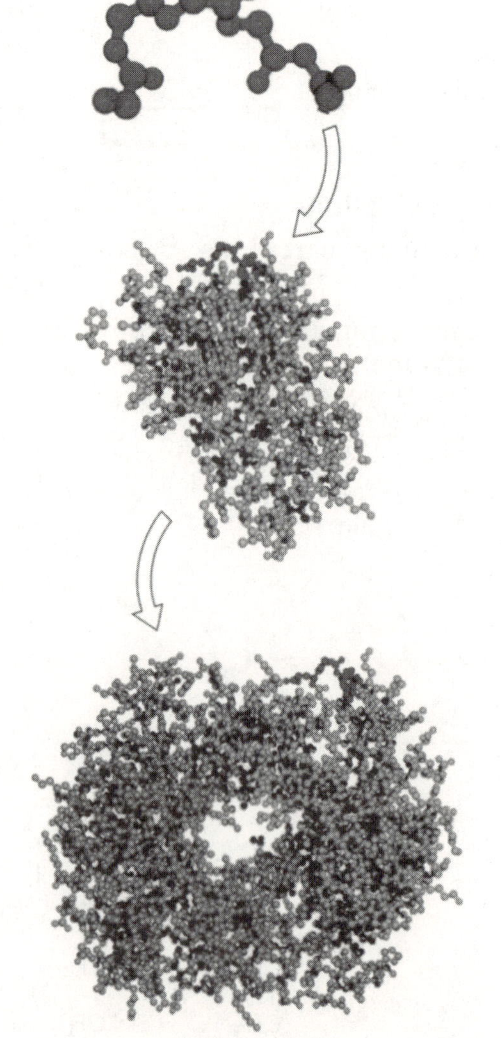

Information:

Secondary structures are normally held in place by *hydrogen bonding*. Two examples of secondary structure are shown in Figure 2.

Figure 2: The α helix (a) and the β sheet (b) are secondary structures of a protein

(a)

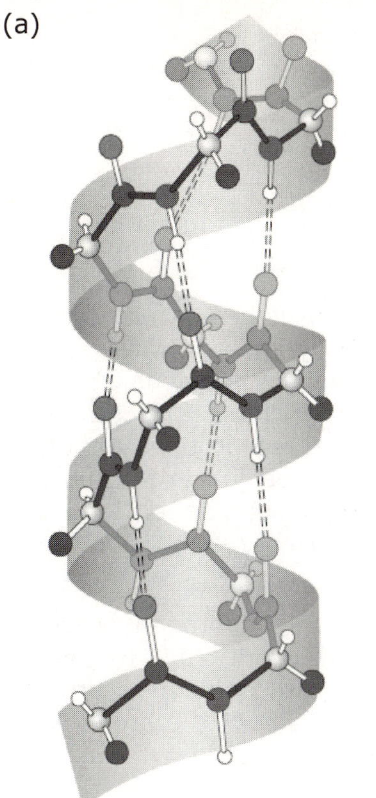

In the α helix, the polypeptide backbone twists so that hydrogen bonds (dashed lines) form between a C=O and an N–H group four residues further along in the sequence.

(b)

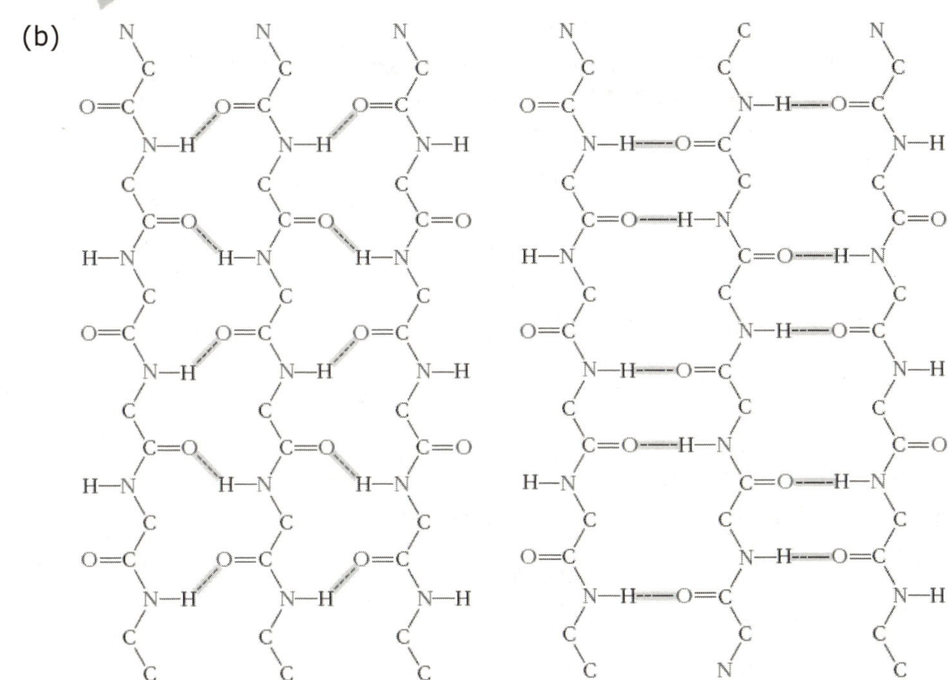

In β sheets, the polypeptide backbone is extended, and hydrogen bonds form between the C=O and N–H groups on adjacent strands. In parallel sheets, the primary sequence of both strands proceeds in the same direction. In antiparallel sheets, the primary sequences proceed in opposite directions.

Critical Thinking Question:

11. Describe how the conformation of the primary structure of a protein is affected when it adopts a particular secondary structure.

Information: Globular proteins have hydrophobic cores

An active protein must be in its *native conformation*. The primary structure is composed of amino acids connected with covalent peptide bonds that are not easily broken. However, the secondary, tertiary, and, quaternary structures are held together with weaker, noncovalent interactions. These can be disturbed by heat, agitation, drastic pH changes, detergents, salts, organic solvents, *etc.*, in a process called **denaturing**. Denatured proteins are often said to be **unfolded**, and are no longer able to perform their intended function. (You can see this visually by frying an egg. As the proteins denature, they become opaque. Good luck hatching a hard-boiled egg!) Some proteins can spontaneously refold, but most denaturations are essentially irreversible, inactivating the protein.

Secondary structures (α helix, β sheet) are held together with only hydrogen bonding. Tertiary and quaternary structures also make use of hydrogen bonding, but the **most important** force holding globular proteins in place is the **hydrophobic effect—**a process by which the water solvent attracts the polar amino acids to the outside of the protein and "squeezes" the nonpolar amino acids into the center of the structure.

Critical Thinking Questions:

12. About half of the 223 amino acid residues of the digestive enzyme trypsin are hydrophobic. Describe where in the tertiary structure you would expect to find those residues.

13. Describe where in the tertiary structure you would expect the polar amino acid residues in trypsin to be located.

14. Another digestive enzyme, chymotrypsin, has two polypeptide chains, each with 245 amino acid residues. Again, about half of the residues have nonpolar side chains. The quaternary structure is shown below (note the two separate polypeptide chains). Describe three distinct locations in the overall structure you would expect the nonpolar amino acid residues to be located.

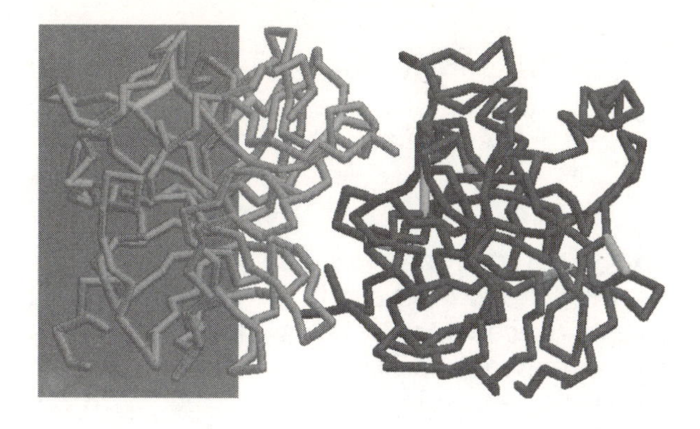

Exercises:

1. Explain why a peptide bond may also be called an amide bond.

2. Rank the solubility of the following amino acids in water at pH 7, from most to least soluble: Val, Ser, Phe, Lys. Explain your answer.

3. Suppose a polypeptide containing 150 amino-acid residues (chosen at random from the 20 common ones) is synthesized in the laboratory. Why is it not correct to call this polypeptide a protein?

4. A detergent, sodium dodecyl sulfate, is shown at the right. Label the hydrophilic and hydrophobic parts of the detergent. Which class of amino acid side chains would be attracted to the hydrophobic part? _______________ Considering your answers to CTQs 12 and 14, explain how adding a detergent to a protein solution might cause the protein to denature.

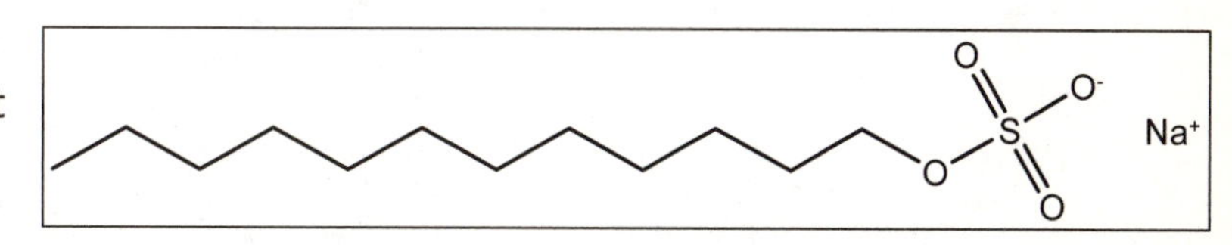

5. Read the assigned pages in your text, and work the assigned problems.

Energy and Metabolism

(What are the energy currencies of the cell?)

Definitions:

metabolism – all biochemical reactions involving the use, production, and storage of energy
anabolism – synthetic (reductive) metabolic reactions which require energy
catabolism – degradative (oxidative) metabolic reactions which produce energy

Information:

Metabolic processes are all those related to the production and use of energy. In the broadest definition, this includes all biochemical reactions of the cell, as all reactions either require or produce energy. We will mostly concern ourselves with catabolic reactions—those which provide the energy required for life (anabolic) processes. Non-photosynthetic organisms such as humans must find all of their required energy in their diet, and so the first stage of energy production begins with digestion. The main classes of food molecules are carbohydrates, proteins, and lipids.

All complex carbohydrates, proteins, and lipids must be broken down into their components in order to be absorbed into the bloodstream. Enzymes in the digestive tract catalyze their breakdown into simple sugars, amino acids and small peptides, and fatty acids. These molecules make their way into the cells *via* the lymphatic and circulatory systems. Once inside cells, they can be broken down (oxidized) and their energy stored.

Table 1: Enzymes that catalyze the hydrolysis of food molecules, and their locations and products

Substrate	Enzymes	Location	Products
complex carbohydrates	amylase, glycosidases	saliva, small intestine	simple sugars and disaccharides
proteins	peptidases, proteases	stomach, small intestine	amino acids and short peptides
lipids	lipases	small intestine	fatty acids and monoacylglycerols

Critical Thinking Question:

1. What type(s) of enzymes would be involved in beginning digestion of the following food sources? Where in the digestive tract would this occur?
    a. Potato starch
    b. Corn oil
    c. Soy protein
    d. Shortening from a pie crust

Information:

Once food molecules have been digested and enter the cells, they enter catabolism. Catabolic pathways are those that involve the breakdown and oxidation of food molecules to produce energy. This process is often called **cellular respiration**.

The energy produced in catabolic reactions is stored in many different molecules, but the most important of these is ATP, **a**denosine **tri**phosphate. It is said that an organism must produce and use its weight in ATP each day.

Under standard conditions, the hydrolysis of ATP to make ADP and water releases 7.3 kcal/mol of energy. Since the reaction is reversible, this means that 7.3 kcal/mol of energy from food molecules are **required** to **make** ATP from ADP and water.

Figure 2: "High-energy" phosphate bonds (indicated in bold) in ATP can be hydrolyzed to release energy.

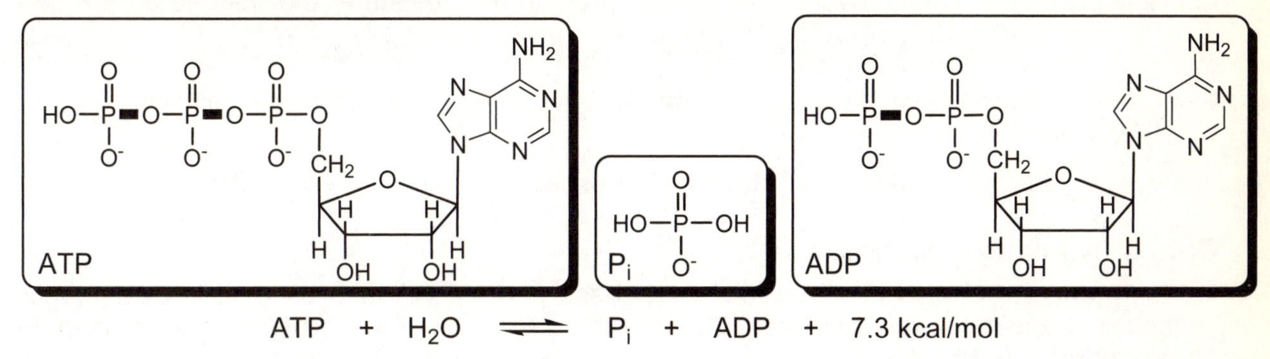

Note the abbreviations in Figure 2 and what they mean: ATP is an adenosine with three phosphates connected together *via* phosphoanhydride bonds; ADP is an adenosine with two phosphates connected together *via* a phosphoanhydride bond; and P_i means *inorganic phosphate*, shown as dihydrogenphosphate ($H_2PO_3^-$). You will sometimes see Ⓟ (a P in a circle) as a shortcut to designate a phosphate group.

It is important to note that energy is **not** released by **breaking** the P-O bond, but by the **making** of the new O-H and P-O bonds during hydrolysis (among other things we need not concern ourselves with). This is why the term "high-energy" bond is shown in quotation marks.

Critical Thinking Questions:

2. The high energy phosphate bond in ADP can be hydrolyzed to release another 7.3 kcal/mol of energy, as shown:

   ADP + H_2O qwe AMP + P_i + 7.3 kcal/mol

   By analogy to Figure 2, draw the two products of this hydrolysis.

3. If an ATP molecule is hydrolyzed (by two water molecules) all the way to AMP (and two P_i) in two steps, how many total kcal/mol of energy could be released? Explain.

Model: ATP is the energy currency of the cell

Consider the phosphorylation of glucose reaction below. The reaction is unfavorable (endothermic) because 3.3 kcal/mol of energy are required.

Glucose + P_i + 3.3 kcal/mol qwe glucose-6-phosphate + H_2O (1)

If ATP is present, it can be hydrolyzed to both provide the needed P_i and the energy.

ATP + H_2O qwe ADP + P_i + 7.3 kcal/mol (2)

When these two reactions are coupled (added) together, the net reaction is favorable (exothermic):

Glucose + ATP qwe glucose-6-phosphate + ADP + 4.0 kcal/mol (3)

Critical Thinking Questions:

4. Copy **all** of the reactants from **both** reactions (1) and (2) in the Model into the space below. Repeat for the products. Confirm that what you have written is equivalent to reaction (3) in the Model.

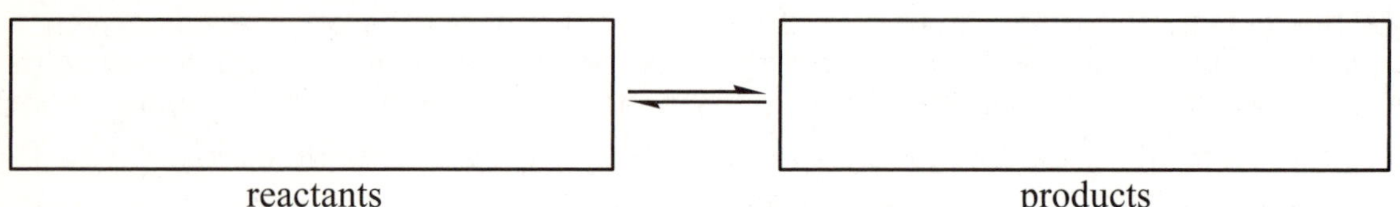

5. Consider the reaction below:

   fructose-6-phosphate + P_i + 3.4 kcal/mol qwe fructose-1,6-bisphosphate + H_2O

   a. In this reaction, is energy **required**, or is energy **released** (circle one)?
   b. If energy from ATP is required for this reaction, in order to provide sufficient energy, would the product of ATP hydrolysis need to be **ADP** or **AMP** (circle one)? Explain your choice.

6. Consider the reaction below:

   phosphoenolpyruvate + H_2O qwe pyruvate + P_i + 14.8 kcal/mol

   a. In this reaction, is energy **required**, or is energy **released** (circle one)?
   b. Would energy from ATP be required for this reaction?
   c. Is sufficient energy produced by this reaction to make ATP from ADP + P_i? Explain.

Information: Other energy currencies

We have defined oxidation as the loss of electrons, and we recognize that the addition of oxygen atoms or loss of hydrogen atoms are signs of oxidation. Since catabolic reactions are mainly oxidative, then **oxidizing agents** (*i. e.,* **electron acceptors**) must be present. The ultimate oxidizing agent in metabolism is O_2, but intermediate steps require coenzymes such as FAD and NAD^+ to accept electrons. The reduced forms of these cofactors represent another location ("currency") in which energy may be stored for later release.

Figure 4: Reduced and oxidized forms of NAD^+. NAD^+ (formed from vitamin B_3, niacin) is involved in oxidations producing C=O double bonds.

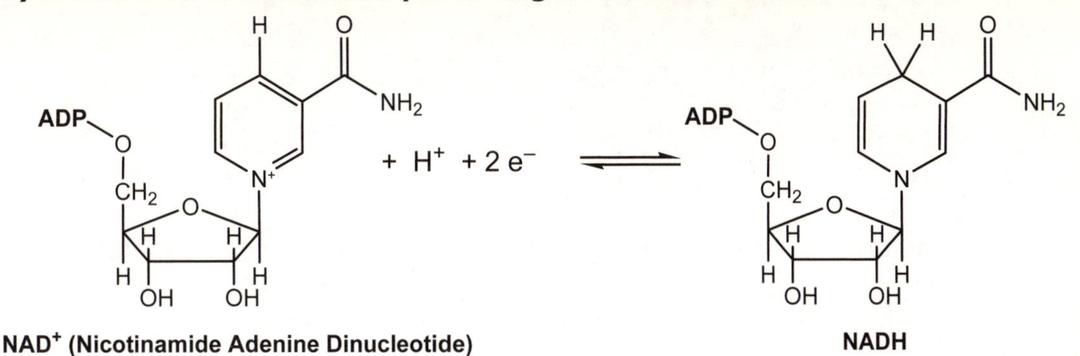

NAD^+ (Nicotinamide Adenine Dinucleotide) **NADH**

For example, ethanol is oxidized to acetaldehyde in liver cells according to the reaction:

$$CH_3CH_2OH + NAD^+ \xrightarrow{\text{alcohol dehydrogenase}} CH_3CHO + NADH + H^+$$

Critical Thinking Question:

7. Consider Figure 4, including the example reaction at the bottom.
   a. Why is the reaction of NAD^+ to NADH and H^+ called a **reduction** of NAD^+?

   b. Due to the shorthand notation for the aldehyde in the example reaction of Figure 4, the location of the C=O that is produced in the reaction is not obvious. Draw a chemical structure that shows the carbonyl group.

Figure 5: Reduced and oxidized forms of FAD; ADP = adenosine diphosphate. FAD (formed from vitamin B_2, riboflavin) is involved in reactions that produce a C=C double bond.

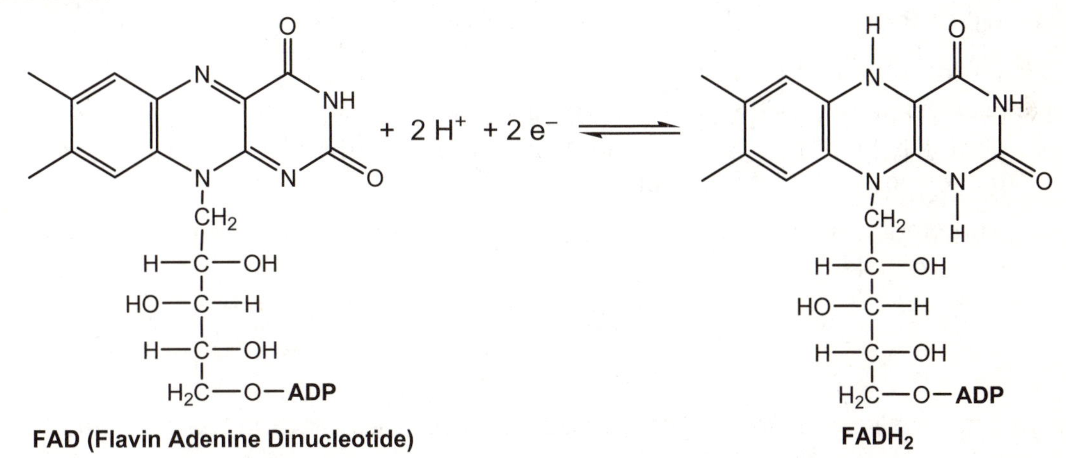

FAD (Flavin Adenine Dinucleotide) **$FADH_2$**

For example, succinate is converted to fumarate by the following reaction:

$$^-O_2C\text{-}CH_2\text{-}CH_2\text{-}CO_2^- + FAD \xrightarrow{\text{succinate dehydrogenase}} {}^-O_2C\text{-}CH{=}CH\text{-}CO_2^- + FADH_2$$

Critical Thinking Questions:

8. Where are the two hydrogen atoms added to FAD? Circle them in the structure of $FADH_2$ in Figure 5.

9. What cofactor (NAD^+, NADH, FAD, $FADH_2$) would be required in the following reaction? Explain.

$$\underset{\text{pyruvate}}{H_3C-\overset{O}{\overset{\|}{C}}-\overset{O}{\overset{\|}{C}}-O^-} \xrightleftharpoons{\text{lactate dehydrogenase}} \underset{\text{lactate}}{H_3C-\overset{OH}{\overset{|}{CH}}-\overset{O}{\overset{\|}{C}}-O^-}$$

10. Which is in the more oxidized form—pyruvate or lactate? Explain how you can tell.

Exercises:

1. Considering the reactions in the Model and your answer to CTQ 4, write a sentence that explains how coupling an unfavorable reaction with ATP hydrolysis can make the reaction favorable.

2. Redraw the example reaction of Figure 5 without the shorthand (that is, use chemical structures that explicitly show the locations of the C=O in the carboxyl groups.

3. Draw a complete balanced reaction for the conversion of pyruvate to lactate (CTQ 8), showing all reactants, products, enzymes and cofactors.

4. We saw in this activity that oxidation reactions of alcohol groups in carbohydrates can provide enough energy to transfer electrons to NAD^+.
   a. Which form of this cofactor is the more oxidized form—NAD^+ or NADH?
   b. Which form is the more reduced form?
   c. Given that in an oxygen-containing environment, oxidation reactions are generally favorable (spontaneous), which form (**oxidized** or **reduced**) is at a higher potential energy level?
   d. In general, would *oxidized cofactors* or *reduced cofactors* provide a "stockpile" of energy? Explain.

5. The enzyme nucleoside *diphosphate kinase* catalyzes the following reaction, with an equilibrium constant, K_{eq}, equal to 1:

   GTP + ADP qwe GDP + ATP

   Explain why ATP and GTP "store" equivalent amounts of energy.

6. Read the assigned pages in your textbook and work the assigned problems.

Enzymes

(Why are biochemical reactions so fast?)

Information:

Catalysts increase the rate of a chemical reaction without being changed themselves. Most biological catalysts are protein **enzymes** that change the way a reaction takes place so that it occurs faster. The reactants in enzyme-catalyzed reactions are called **substrates**.

Enzymes lower the activation energy of a reaction by binding one or more substrates into an **active site**, using hydrophobic or hydrophilic interactions, hydrogen bonding, *etc.* This binding stretches each substrate into a reactive form and aligns it properly for the chemical reaction to take place.

Figure 1 (at right): Tertiary structure of the digestive enzyme trypsin from *Streptomyces griseus*. The protein substrate binds into the active site, indicated with darker color.

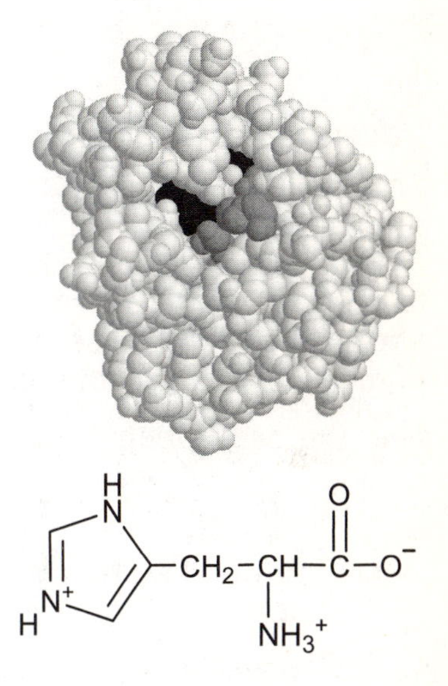

Critical Thinking Questions:

1. Circle the side chains in each amino acid below. To which class (hydrophobic, polar, charged) does each belong? (You should be able to do this without looking at a table).

$^{-}O-C(=O)-CH_2\cdot CH(NH_3^{+})-C(=O)-O^{-}$ $\qquad CH_3\cdot CH(NH_3^{+})-C(=O)-O^{-}$ $\qquad HO-CH_2\cdot CH(NH_3^{+})-C(=O)-O^{-}$ $\qquad (C_3H_3N_2H_2^{+})-CH_2-CH(NH_3^{+})-C(=O)-O^{-}$

2. Suppose that the first amino acid (on the left) in CTQ 1 is part of a substrate. Hypothesize on which of the other three amino acid residues might be present in the active site of an enzyme in order to bind to the side chain of the substrate. Explain.

Model 1: Energy diagrams for uncatalyzed and catalyzed exothermic reactions

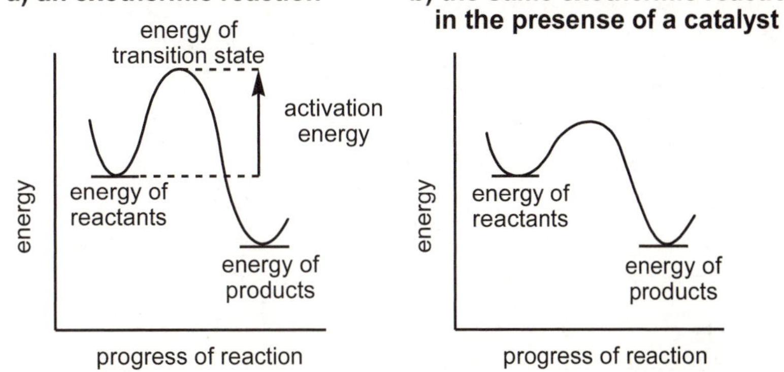

Critical Thinking Questions:

3. Does an enzyme change the energies of the reactants or products?
4. Considering your answer to CTQ 3, does an enzyme change the *equilibrium amounts* of reactants and products? Explain.

5. Draw a vertical arrow onto Model 1 (b) that represents the magnitude of the activation energy.
6. Explain the function of an enzyme in a complete sentence or two.

Information: Six classes of enzymes

Enzymes are classified according to the type of reaction they catalyze. These reactions are often divided into six basic types. Enzymes are often named after their substrate and reaction class, and have the suffix "-ase." For example, the enzyme *triose phosphate isomerase* catalyzes the isomerization of glyceraldehyde 3-phosphate and dihydroxyacetone phosphate—two "triose phosphates." Since all reactions are reversible, the enzyme may be named according to the reverse of the normal reaction, *i. e.,* using the name of a product.

Find the description of the six classes of enzymes in your textbook. You are responsible for identifying reactions in the six classes. Here are some tricks to help you keep the six classes straight:

1. **oxidoreductase**: catalyzes a redox reaction; adds oxygen or removes 2 hydrogen atoms from substrate; requires a cofactor such as NAD^+, $NADP^+$ or FAD;
2. **transferase**: transfers a functional group (*e. g.,* NH_2 or phosphate) between substrates;
3. **hydrolase**: catalyzes a hydrolysis reaction; substrate + $H_2O \rightarrow$ two products;
4. **lyase**: adds or remove groups involving a double bond (no ATP required);
5. **isomerase**: makes an isomer of the substrate by rearrangement;
6. **ligase**: forms a bond to join two substrates using ATP hydrolysis for energy.

Critical Thinking Questions:

7. Which of the six classes of enzymes catalyze each of the following reactions?

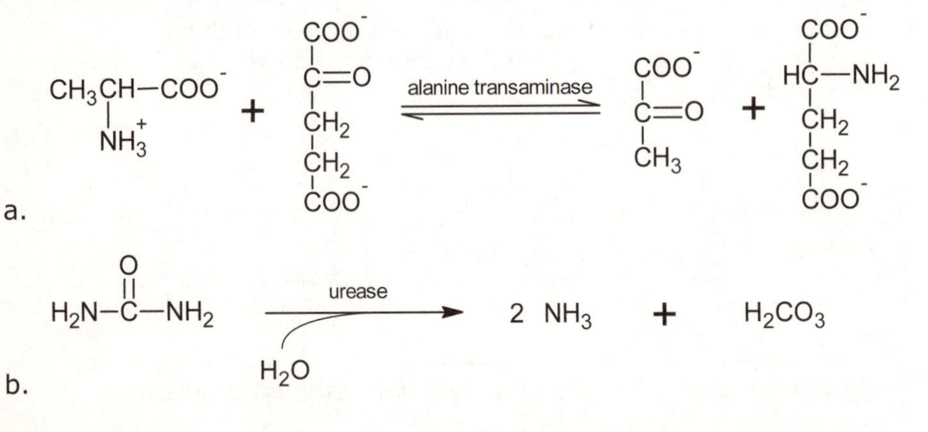

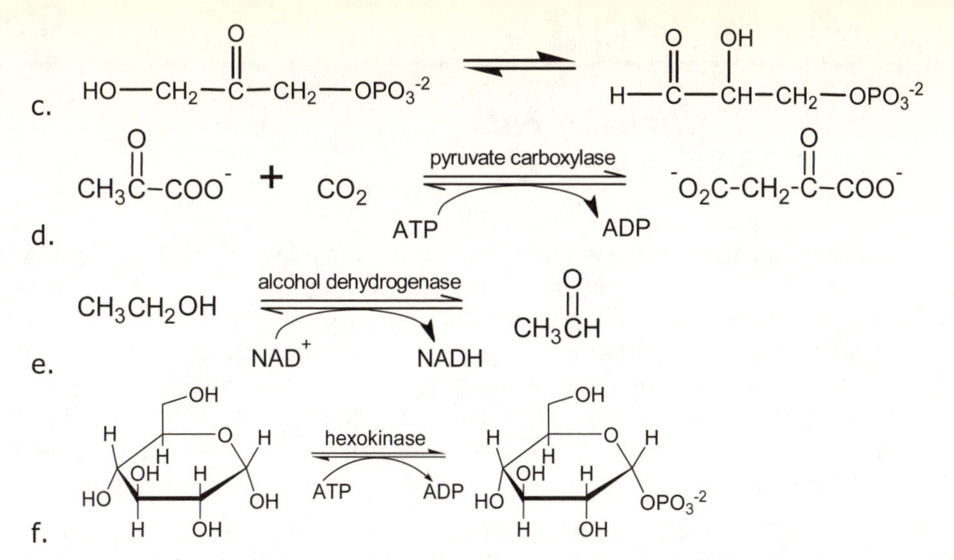

8. What is the name of the substrate molecule shown for hexokinase in CTQ 7f? Be specific.

Information: Inhibition of enzymes

Many pharmaceuticals are enzyme inhibitors. These inhibitors may be either **competitive**, if they bind to the active site to keep the substrate from entering, or **noncompetitive**, if they bind elsewhere to the enzyme and inactivate it by changing its shape.

An example is a treatment for the disease emphysema. The enzyme **elastase** is a protease which helps to maintain connective tissue in the lungs, and elsewhere. It does this by catalyzing the breakdown of old proteins in these tissues. Emphysema is a lung disease caused when the body's ability to produce a natural inhibitor of elastase is compromised. When this natural inhibitor is absent, the elastase gets out of control and destroys large amounts of healthy tissue.

Prolonged study by chemists has enabled the synthesis of a molecule that resembles the shape of the natural inhibitor, and can be given in an inhaler to those suffering from emphysema. This molecule is a competitive inhibitor of elastase, and slows its action by filling up the active site and preventing it from binding to (and hydrolyzing) the proteins in the lungs.

Critical Thinking Question:

9. Glucosamine (shown below) is an inhibitor of hexokinase. Would you expect it to be competitive or noncompetitive? Explain.

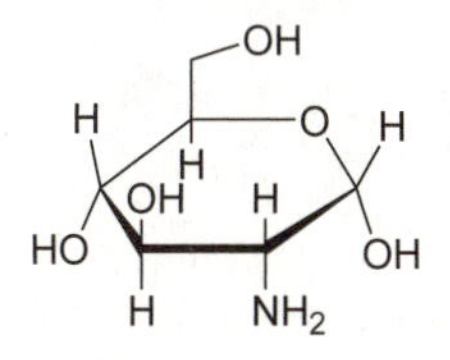

Exercises:

1. Often, an enzyme requires a cofactor or prosthetic group in its active site in order to be an efficient catalyst. Many such cofactors are synthesized by plants and obtained in the diet as vitamins. Find a description in your text of a particular B vitamin. Give its name, and describe its biological function in terms of the name(s) and function(s) of the enzyme(s) that require it.

2. Read the assigned pages in your textbook and work the assigned problems.

Nucleic Acids

(What is DNA made of?)

Information:

Nucleic acids were so named because they were acidic molecules found in the nucleus of atoms. Like other biomolecules such as polysaccharides and proteins, nucleic acids are polymers made from a small number of building blocks. The two main types of nucleic acids are **r**ibo**n**ucleic **a**cid (RNA) and 2′-**d**eoxy**r**ibonucleic **a**cid (DNA). DNA contains the genetic material of a cell, and is found in the nucleus. (The chromosomes which form during cell replication contain all the cellular DNA, along with proteins.) During the life of a cell, the information in DNA is copied to RNA, which contains the information needed to synthesize each protein that the cell requires.

The building blocks of RNA and DNA are nucleotides (See Figure 1). Note that the carbons in the sugar residue are numbered with "primes"—1′, 2′, 3′, 4′, and 5′ (read one prime, two prime, *etc.*) to distinguish them from the numbers of the carbons in the base. RNA contains the sugar ribose, which has hydroxyl (-OH) groups at the 2′ and 3′ positions. In DNA, the 2′ hydroxyl group is missing.

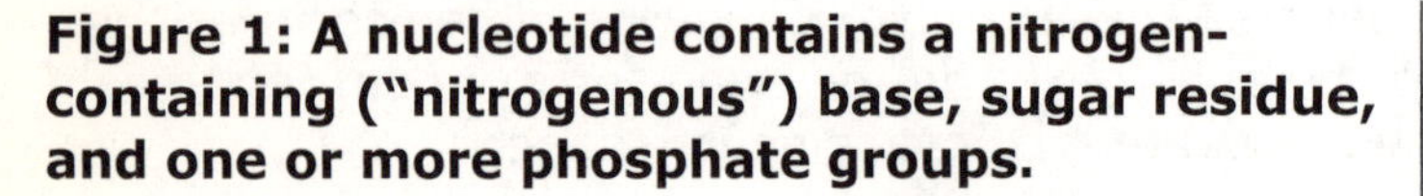

Figure 1: A nucleotide contains a nitrogen-containing ("nitrogenous") base, sugar residue, and one or more phosphate groups.

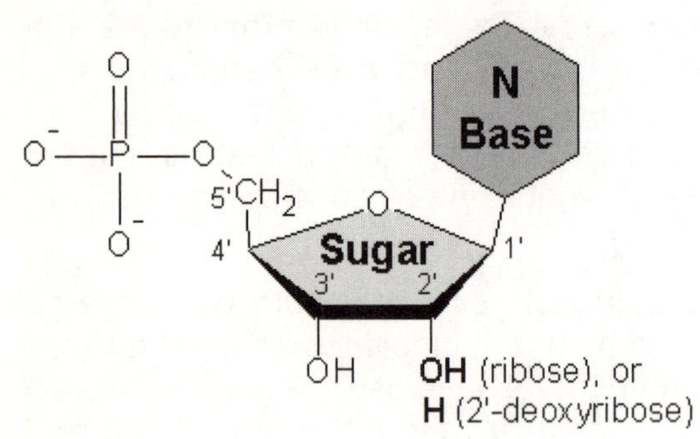

Figure 2: The organic molecules (bases) pyrimidine and purine

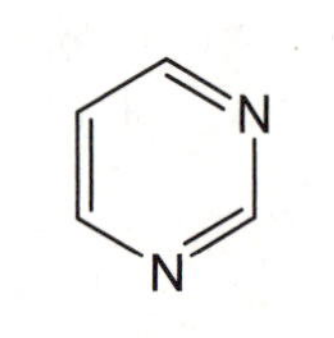

pyrimidine

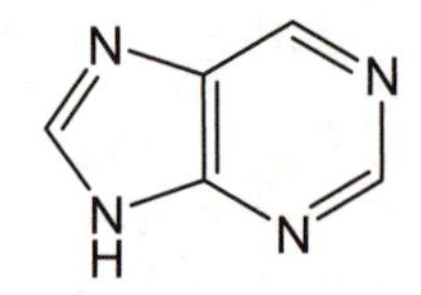

purine

Figure 3: The nitrogen-containing bases in DNA and RNA

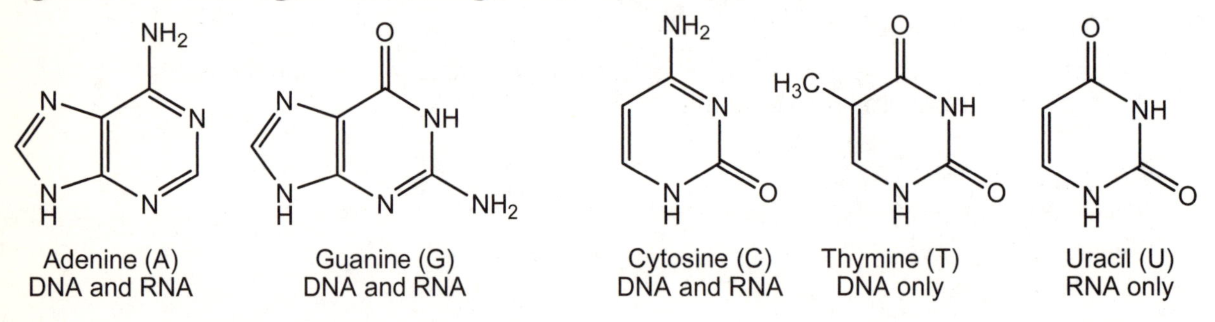

Critical Thinking Question:

1. Based on Figures 2 and 3, which nitrogen-containing bases in DNA and RNA are purines? Which are pyrimidines?

Information:

A nucleoside is a nitrogen-containing base connected to a sugar (ribose or 2′-deoxyribose) residue. A nucleotide is a nitrogen-containing base connected to a sugar residue **and** at least one phosphate group.

Memory device: "A nucleo**tide** has a phosphate **tied** on."

Table 1: Names of nucleosides and some nucleotides

RNA Base	Nucleoside	Nucleotide
Adenine (A)	Adenosine (A)	Adenosine-5′-monophosphate (AMP)
Guanine (G)	Guanosine (G)	Guanosine-5′-diphosphate (GDP)
Cytosine (C)	Cytidine (C)	Cytidine-5′-triphosphate (CTP)
Uracil (U)	Uridine (U)	Uridine-5′-monophosphate (UMP)
DNA Base	Nucleoside	Nucleotide
Adenine (A)	Deoxyadenosine (A or dA)	Deoxyadenosine-5′-monophosphate (dAMP)
Guanine (G)	Deoxyguanosine (G or dG)	Deoxyguanosine-5′-diphosphate (dGDP)
Cytosine (C)	Deoxycytidine (C or dC)	Deoxycytidine-5′-triphosphate (dCTP)
Thymine (T)	Deoxythymidine (T or dT)	Deoxythymidine-5-monophosphate (dTMP)

Critical Thinking Question:

2. Refer to Table 1. Write the full name of the molecule with the following abbreviations.

   a. ATP ____

   b. UDP ____

   c. dA ____

Information: Levels of structure of nucleic acids

Similar to the levels of protein structure that we have seen, nucleic acids also have primary, secondary, and tertiary structures. The **primary structure** is the nucleic acid sequence. The nucleosides are connected from the 5′ phosphate on one nucleoside to the 3′ hydroxyl group on the next, *via* a **phosphodiester** linkage (see Figure 4).

Figure 4: Nucleotides condense together *via* phosphodiester linkages.

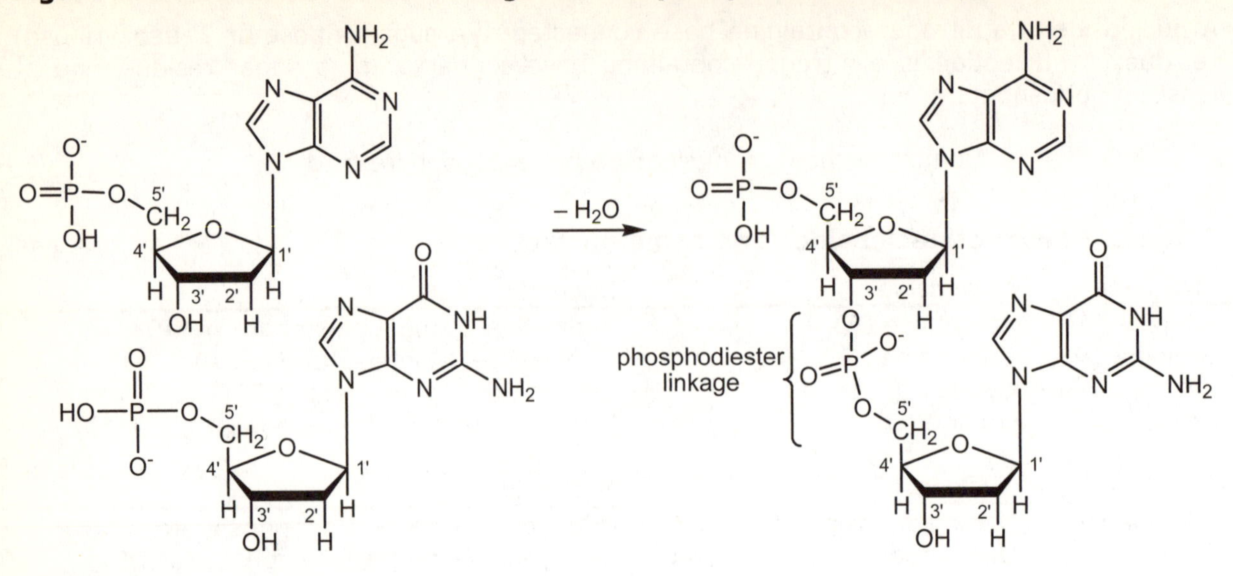

Critical Thinking Questions:

3. Is the dinucleotide in Figure 4 RNA or DNA? Explain how you can tell.

4. The 5′ end of a polynucleotide has a free 5′-phosphate, and the 3′ end has a free 3′-hydroxyl group. Name the nucleotide at the 5′ end of the dinucleotide in Figure 4.

Information:

When nucleosides are connected together, there is only one free 5′ phosphate and one free 3′ hydroxyl group. By convention, the sequences (primary structures) of DNA and RNA are always given from the 5′ end to the 3′ end. Therefore, AAAAGT is not the same as TGAAAA.

The most important **secondary structure** of DNA is the famous double helix. In this structure, two sugar-phosphate backbones spiral around each other, and are held together by **hydrogen bonds** between pairs of nitrogen bases, called **base pairs**. Figure 5 illustrates how the hydrogen bonding partnership works. Note that bases almost never pair with the wrong partner, because the hydrogen bonds would not line up correctly.

Figure 5: Hydrogen bonds between AT and CG (complementary) base pairs hold the DNA double helix together.

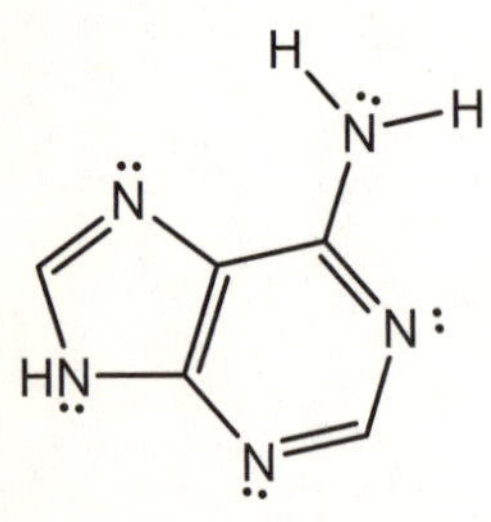

Adenine (A)

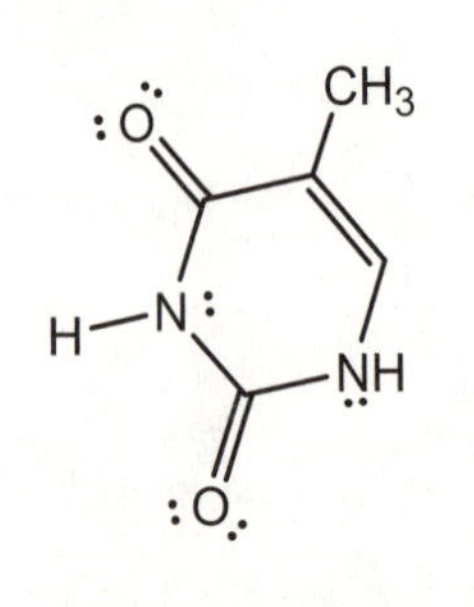

Thymine (T)

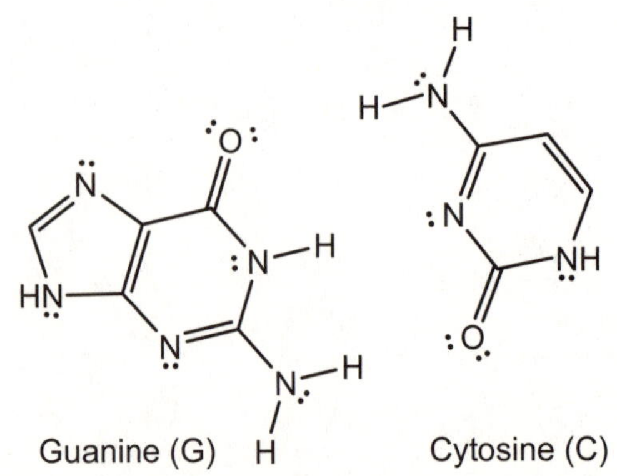

Guanine (G) Cytosine (C)

Critical Thinking Questions:

5. Review: In general, two conditions are required for hydrogen bonding to occur:
   a. There must be a hydrogen atom bonded to a ________________ or ________________ atom (to *donate* the hydrogen bond).
   b. There must be a ____ (to *accept* the hydrogen bond).
6. Use dashed lines to draw hydrogen bonds between A and T in the appropriate locations in Figure 5. Do the same for the CG pair.
7. DNA double helices with more CG content are more heat stable than those with more AT content. Explain why this is so. (See Figure 5.)
8. Words or phrases that read the same forwards and backwards are called palindromes. Examples are "radar" or the phrase "A man, a plan, a canal: Panama!" Suppose that one strand of DNA is palindromic (such as ATGGTA). Would the complementary strand also be palindromic? (Hint: Wwrite the complementary sequence.) Explain.

Information: Tertiary structure of nucleic acids

Individual strands of RNA or DNA can also have tertiary structure. Consider the typical RNA molecule shown in Figure 6.

Figure 6: The secondary (a) and tertiary (b) structures of a typical transfer RNA molecule. The three-base *anticodon* hydrogen-bonds to a messenger RNA *codon*, and the free 3′ end forms an ester bond to a particular amino acid (in this case, phenylalanine).

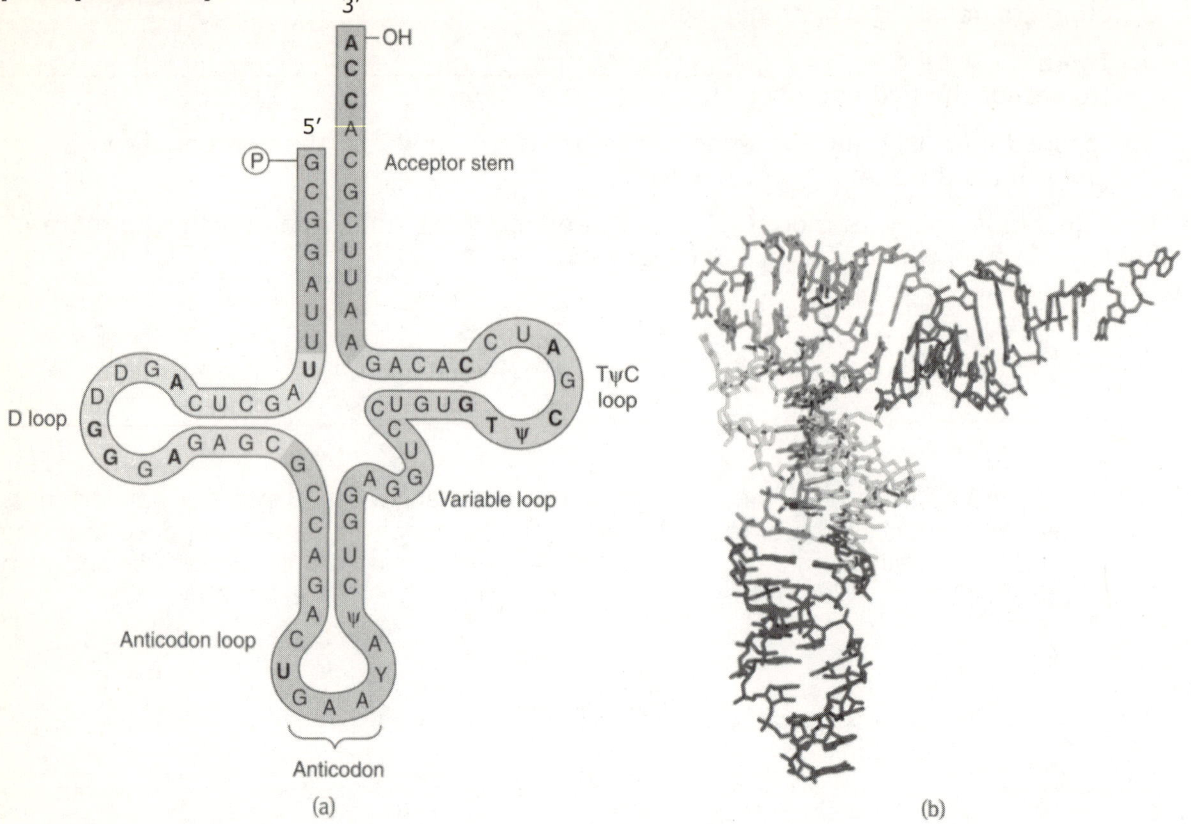

[From Essential Biochemistry; Pratt, C.W. and Cornely, K.; Copyright © (2004) by John Wiley & Sons, Inc. Reprinted with permission of John Wiley & Sons, Inc.]

Critical Thinking Question:

9. RNA is used as the "go-between" to carry information from DNA and incorporate it into proteins. Explain why molecules like transfer RNA (Figure 6) would be necessary in such a process. (Hint: What are the building blocks of proteins?)

10. Consider the transfer RNA in Figure 6b, in which the hydrogen atoms are *not* shown:

    a. Describe where in the tertiary structure you would expect the ribose and phosphate backbone to be.

    b. Ribose is a 5-membered ring. Identify (and circle) at least three ribose rings in Figure 6b. Does their location match your prediction in Exercise 10a?

c. Describe where in the tertiary structure you would expect the base pairs to be.

d. Draw an arrow to at least two locations in Figure 6b where hydrogen bonds between base pairs would be.

Exercises:

1. Identify each species as a nucleoside or nucleotide.

    a. adenine

    b. cytidine

    c. deoxythymidine

    d. UDP

2. Identify each base as a purine or a pyrimidine.

    a. cytosine

    b. hypoxanthine, 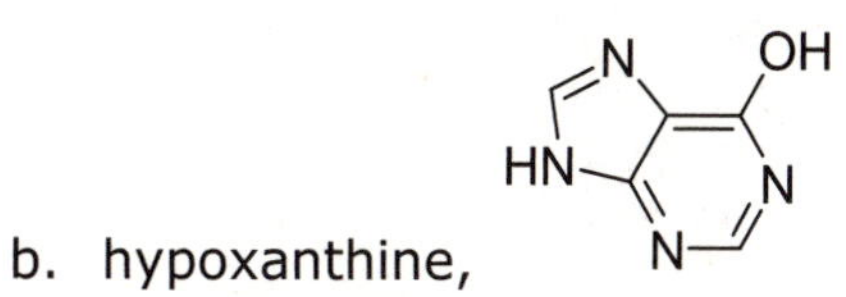

3. Read the assigned pages in your textbook, and work the assigned problems.

Glycolyis

(How is glucose converted to pyruvate?)

Information:

Catabolism is the breakdown of food molecules for energy. This process involves digestion and hydrolysis of carbohydrates, lipids, and proteins; degradation of these molecules into smaller ones in the cytosol of cells; and finally, oxidation of these molecules to produce energy in various metabolic pathways. Initial stages of catabolism occur in the cytosol (cytoplasm), while later stages (and most of the oxidation reactions) occur in mitochondria.

These pathways may be of three types: **linear**, in which the product of one step is the substrate for the next; **cyclic**, which is like linear except that one substrate is regenerated during the cycle so that the steps repeat over again; and **spiral**, in which the steps repeat, but the substrate molecules change slightly each time.

We begin our examination of metabolic pathways with the oxidation of glucose to produce energy. Glucose is metabolized to CO_2 and H_2O *via* three consecutive pathways: **glycolysis** (in the **cytosol**); the **citric acid cycle**—or Krebs' cycle—(in the **mitochondrial matrix**); and the **electron transport chain** (in the **inner mitochondrial membrane**). We will examine each of these three pathways in the next three ChemActivities.

Definitions:

aerobic process – a cellular process which requires oxygen
anaerobic process – a cellular process which does not require oxygen

Critical Thinking Questions:

Refer to Figure 1 (on the following page) to help you answer CTQs 1-11.

1. Which steps of glycolysis are referred to as the "energy investment phase?"

2. **Why** are these steps referred to as the "energy investment phase?"

3. In step 4 of glycolysis, the 6-carbon molecule fructose-1,6-bisphosphate is converted into two 3-carbon molecules—glyceraldehyde-3-phosphate and dihydroxyacetone phosphate. What must happen to the dihydroxyacetone phosphate in order for it to continue proceeding through the glycolytic pathway?

4. Consider steps 6-10 of glycolysis.
    a. Why is there a "2" in front of each substrate in steps 6-10 (*e. g.*, 2 1,3-Bisphosphoglycerate, 2 3-Phosphoglycerate, *etc.*)

    b. Why are these steps referred to as the "energy payoff phase?"

Figure 1: In glycolysis, the six-carbon glucose molecule is converted to two three-carbon pyruvate molecules, producing 2 ATP and 2 NADH molecules.

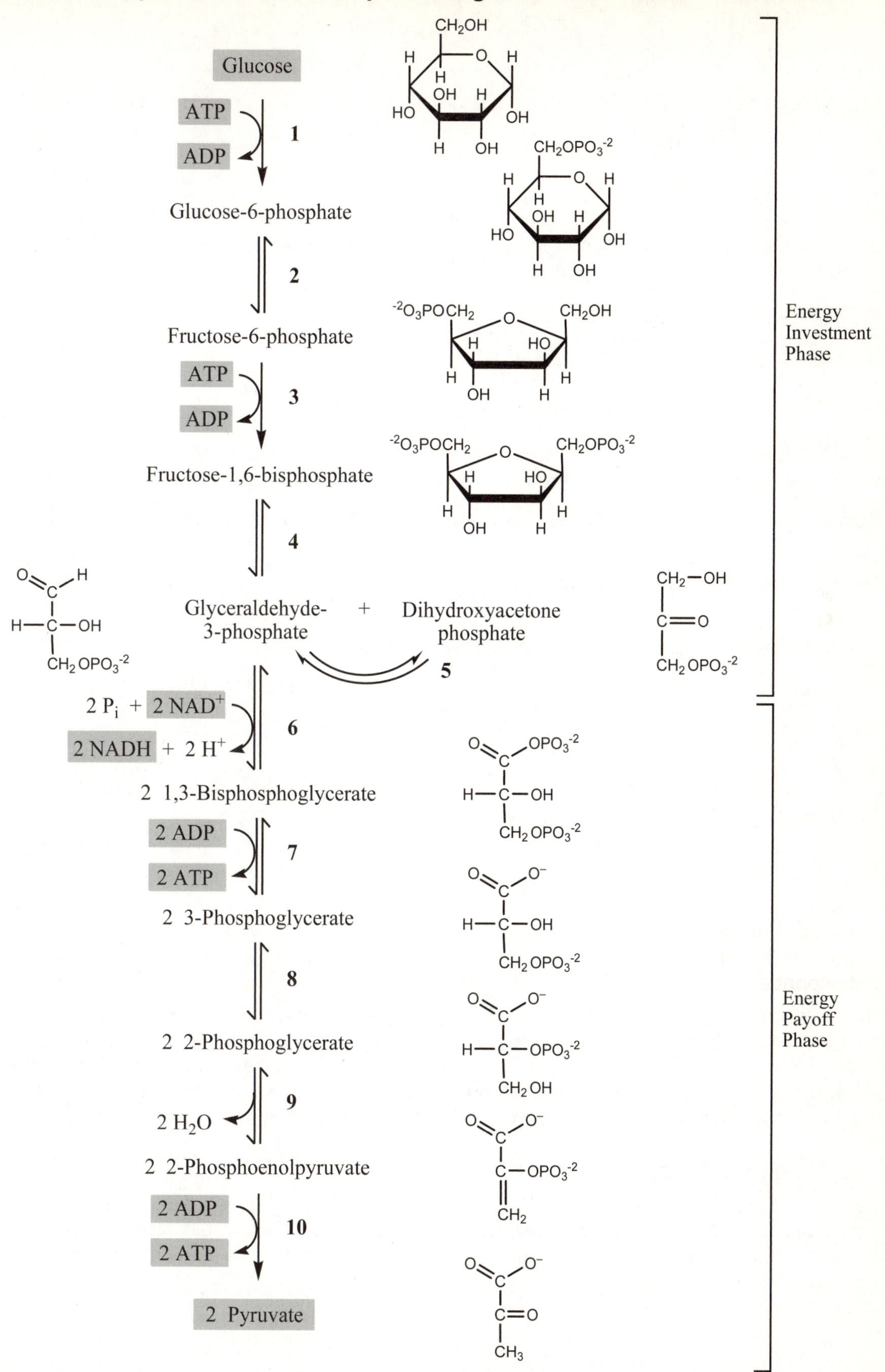

5. ATP is used in two steps (1 and 3) and generated in two steps (7 and 10). Explain, then, how glycolysis gives a net yield of 2 molecules of ATP.

6. Only one of the ten steps of glycolysis is an oxidation. Which one? How can you tell?

7. Using only the information in Figure 1, it is possible to determine the class of enzyme used for each of the ten steps of glycolysis. Which of the six classes of enzymes would be required for the following steps?

   a. Step 3 ______________________

   b. Step 6 ______________________

   c. Step 2 ______________________

   d. Step 8 ______________________

8. Is oxygen required as a substrate for any of the ten steps of glycolysis?

9. Based on your answer to CTQ 8, is glycolysis an aerobic or anaerobic pathway?

10. Erythrocytes (red blood cells) contain no mitochondria. Can erythrocytes obtain energy from glycolysis? Why or why not?

11. The net chemical equation for the first step of glycolysis is:

    Glucose + ATP qwe glucose-6-phosphate + ADP

    Write a similar equation for the following steps of glycolysis:

    a. Step 7:

    b. Step 4:

    c. Step 10:

    d. Step 6:

Information:

Glycolysis does not require oxygen, but it does require an oxidizing agent, NAD^+, in step 6. Under aerobic conditions, the NADH produced during glycolysis is converted back to NAD^+ in the mitochondria, and glycolysis can continue. But under anaerobic conditions, the cell needs another method to regenerate the NAD^+, or else energy production would cease. This method in most organisms is to produce lactate (lactic acid), as shown in Figure 2.

Critical Thinking Questions:

12. Lactic acid is, of course, acidic. When muscles contract, they squeeze the blood out of the vessels, and therefore are operating anaerobically. What would happen to the pH of muscle cells during prolonged contraction? How might this help explain muscle soreness from a workout?

Figure 2: Under anaerobic conditions, pyruvate is converted to lactate. The production of acetyl-SCoA under aerobic conditions is often called the "bridge" or "transition" step to the citric acid cycle.

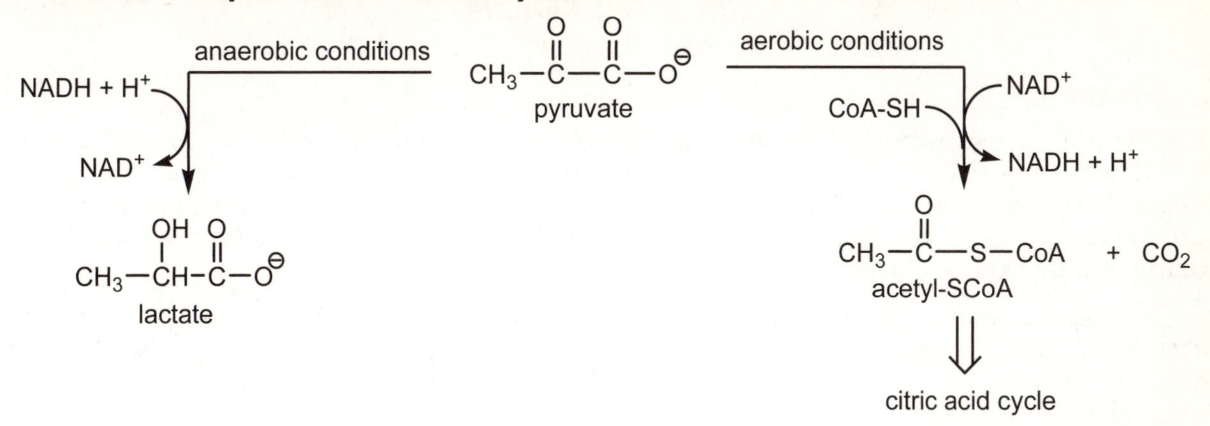

Exercises:

1. Lactate eventually makes its way to the liver, where it is converted back into pyruvate. Explain why lactate production is called a "dead-end" pathway.

2. In the absence of oxygen, yeast cells regenerate NAD^+ by converting pyruvate into ethanol and carbon dioxide. Write a balanced chemical equation for this reaction.

3. Using only the information in Figure 1, it is possible to determine the class of enzyme used for each of the ten steps of glycolysis. Which of the six classes of enzymes would be required for the following steps?

   a. Step 1 ______________________

   b. Step 4 ______________________

   c. Step 5 ______________________

   d. Step 9 ______________________

4. The net chemical equation for the first step of glycolysis is:

   Glucose + ATP qwe glucose-6-phosphate + ADP

   Write a similar equation for the following steps of glycolysis:

   a. Step 3:

   b. Step 5:

   c. Step 8:

   d. Step 9:

5. Read the assigned pages in your textbook and work the assigned problems.

Citric Acid Cycle

(How is pyruvate oxidized to CO_2 and H_2O?)

Information:

Following glycolysis, the next stage of aerobic glucose catabolism is the oxidation of acetyl-CoA to CO_2 and H_2O in the mitochondria, the **citric acid cycle**—also known as the citrate cycle, Krebs' cycle, or TCA (for tricarboxylic acid) cycle. This ChemActivity will focus on the citric acid cycle (CAC). Following the CAC is the pathway including **electron transport** and **oxidative phosphorylation** (ChemActivity 42).

Recall that the end product of aerobic glycolysis is pyruvate, which is then converted to acetyl-CoA (See Figure 2 of ChemActivity 40). The CAC takes the carbons of acetyl-CoA and oxidizes them to CO_2 and H_2O, producing the reduced cofactors NADH and $FADH_2$, and a small amount of ATP.

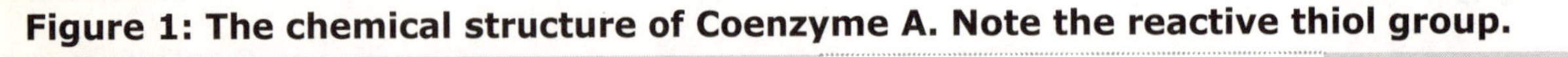

Figure 1: The chemical structure of Coenzyme A. Note the reactive thiol group.

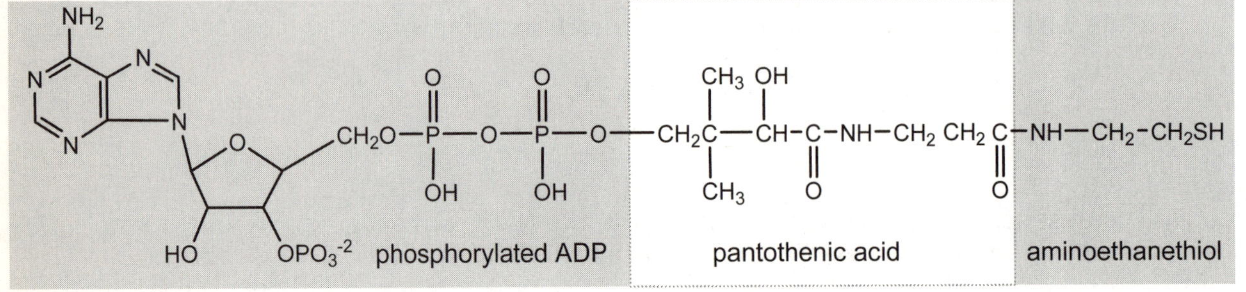

Critical Thinking Question:

1. The part of Coenzyme A that originated in pantothenic acid is actually only an acid *residue*. It is the result of two condensation reactions—with aminoethanethiol on one end, and with the phosphorylated ADP on the other end. Draw the structure of pantothenic acid. (Recall that condensation reactions normally join two molecules and release a **water** molecule.)

Figure 2: The reactive thiol group of Coenzyme A (CoA or CoA-SH) can carry an acetyl group. The resulting molecule is abbreviated "acetyl-CoA" or "acetyl-SCoA"

Under aerobic conditions, the NADH and $FADH_2$ produced in the CAC are oxidized back to NAD^+ and FAD (We will see this in ChemActivity 42). While oxygen is not directly required for the CAC, a lack of oxygen would cause NAD^+ and FAD to be unavailable, and so the CAC would slow to a halt.

Seven of the eight CAC enzymes are found in the mitochondrial matrix. The eighth is bound in the inner mitochondrial membrane.

Figure 2: The citric acid cycle oxidizes acetyl-CoA to CO_2 and H_2O

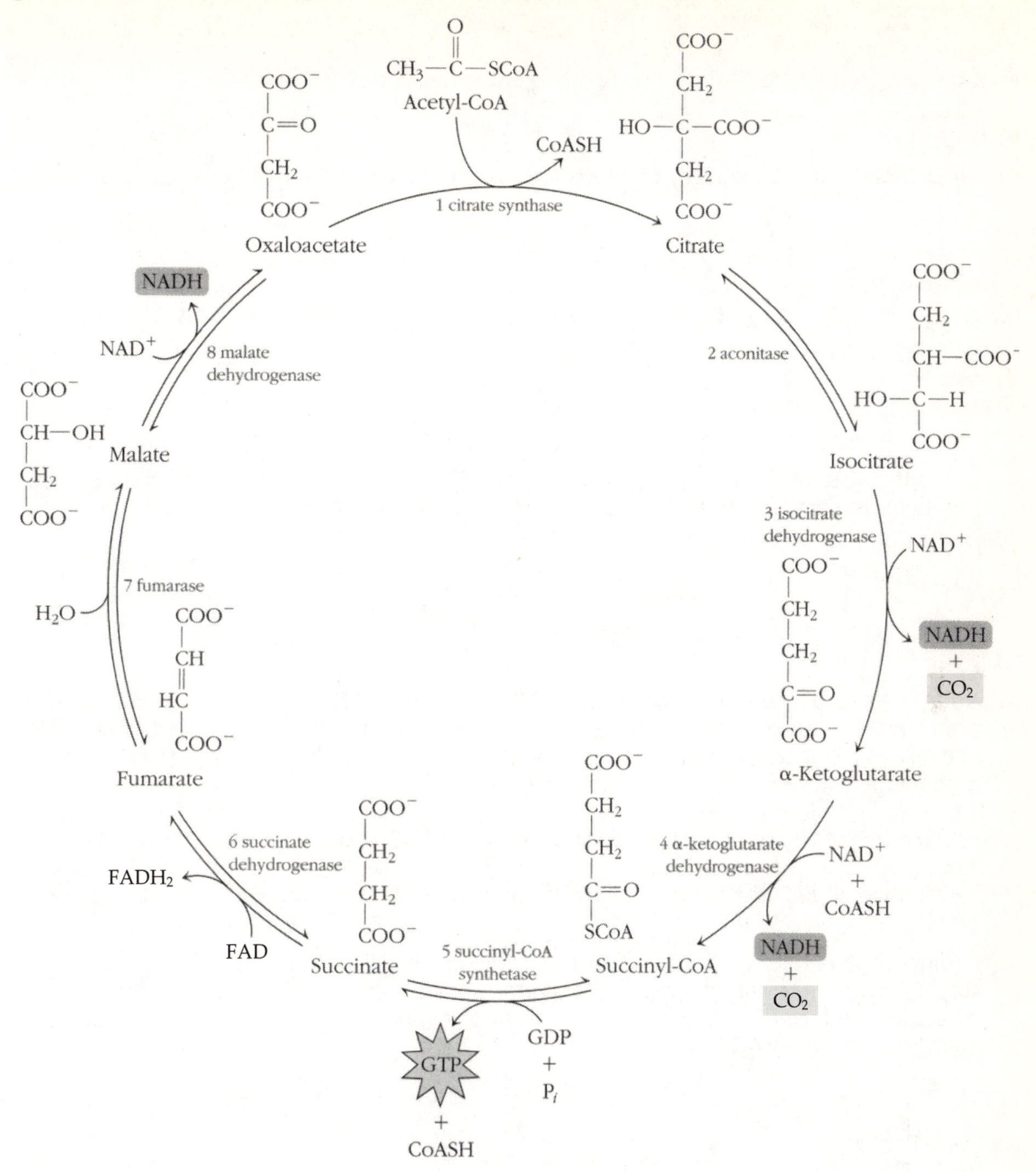

Critical Thinking Questions:

Refer to Figure 2 to help you answer CTQs 1-8.

2. In step 1 of the CAC the acetyl group is transferred from acetyl-CoA to oxaloacetate. Step 8 produces oxaloacetate. Explain why this pathway is said to be a **cyclic** pathway.

3. In each turn of the CAC, two molecules of CO_2 are released. In which steps is CO_2 released?

4. In which step of the CAC do two carbons enter the cycle?

5. What metabolite (substrate) provides the source of carbons for the CAC to oxidize?

6. Four of the eight steps in the CAC are oxidations. Which four? How can you tell?

7. It is difficult to oxidize a tertiary alcohol since the carbon with the hydroxyl group has no hydrogen to remove. What do you think is the purpose of Step 2?

8. No ATP is produced directly by the CAC, but one GTP (guanosine triphosphate) is produced in Step 5. Other "energy-currency" cofactors produced in the CAC are NADH and $FADH_2$. For each turn (8 steps) of the CAC, circle the number of each that are produced.

   Number of NADH produced in one turn of the CAC (circle one) – **1 2 3**

   Number of $FADH_2$ produced in one turn of the CAC (circle one) – **1 2 3**

9. Adding your results from CTQ 8 to the cofactors produced in glycolysis and the bridge step (see ChemActivity 40, Figure 2), total up the number of ATP molecules and other cofactors produced in the entire oxidation of <u>one molecule of glucose</u> to CO_2 and H_2O *via* <u>glycolysis and the CAC</u>.

   ATP ____

   GTP ____

   NADH ____

   $FADH_2$ ____

10. Using only the information in Figure 2, it is possible to determine the class of enzyme used for the eight steps of the CAC. Which of the six classes of enzymes would be required for the following steps?

    a. Step 2 ____________________

    b. Step 4 ____________________

    c. Step 5 ____________________

    d. Step 7 ____________________

Exercises:

1. The chemical equation for the first step of the CAC is:

   Oxaloacetate + acetyl-SCoA + H_2O qwe citrate + CoA-SH

   Write a similar equation for the following steps of the CAC:

   a. Step 2:

   b. Step 4:

   c. Step 5:

   d. Step 7:

2. Describe what cellular respiration is, in general (including when it occurs).

3. Using your textbook as a reference, sketch and label a diagram of a mitochondria (in a human muscle cell, for example).

4. Read the assigned pages in your text, and work the assigned problems.

Electron Transport/Oxidative Phosphorylation

(How can energy from the reduced cofactors be used to make ATP?)

Following the CAC (ChemActivity 41) in the aerobic oxidation of glucose is the final oxidative pathway, which includes **electron transport** and **oxidative phosphorylation**, known also as the **electron transport chain** (ETC).

At the end of the citric acid cycle, all six carbons of glucose have been oxidized to CO_2, and a few nucleotide triphosphates (ATP and GTP) have been produced. However, most of the energy has been saved in the reduced cofactors NADH and $FADH_2$. The ETC takes the electrons from NADH and $FADH_2$ and passes them through a "chain" of many other cofactors and enzyme complexes, until they finally end up adding to O_2 to make H_2O.

Information: The Electron Carriers in the ETC

The proteins and cofactors shown in Figures 1-4 carry electrons within and between the ETC enzyme complexes, either one or two at a time.

Figure 1: The structure of FMN is similar to FAD (ChemActivity 37, Figure 5), but with a phosphate group (at bottom) in place of ADP.

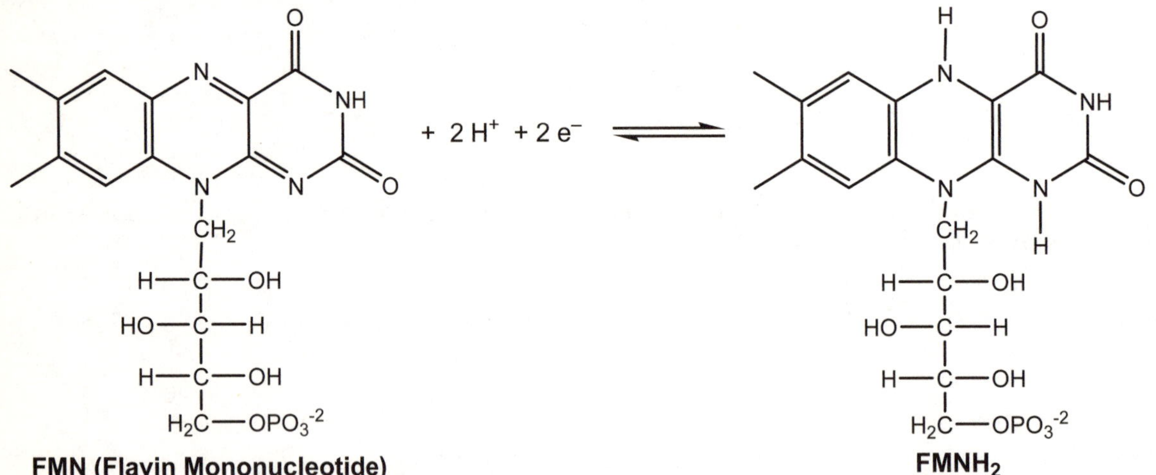

Figure 2: Iron-sulfur proteins are proteins containing *iron-sulfur clusters*—in which the iron is at least partially coordinated by sulfur. A typical iron-sulfur cluster is shown; curved lines indicate continuation of the protein backbone.

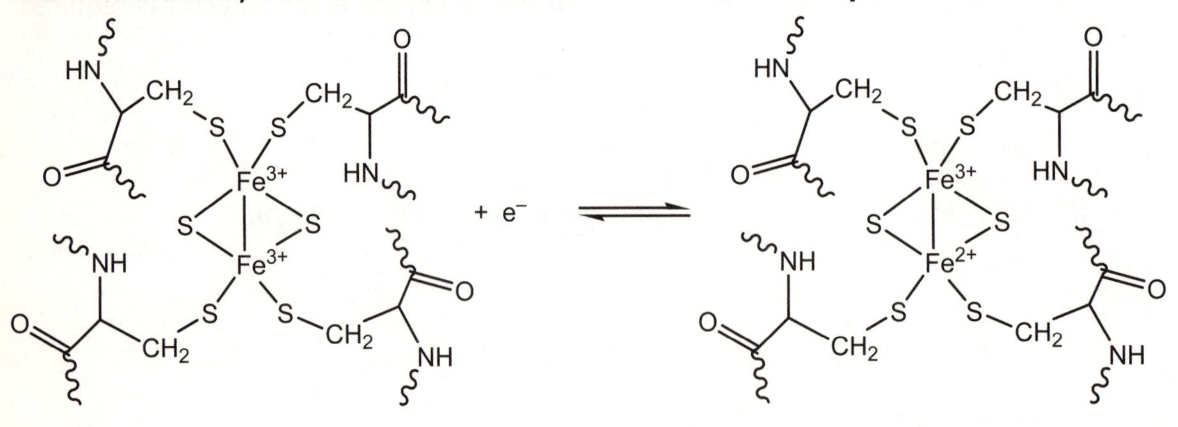

Figure 3: Oxidized and reduced forms of Coenzyme Q (ubiquinone)

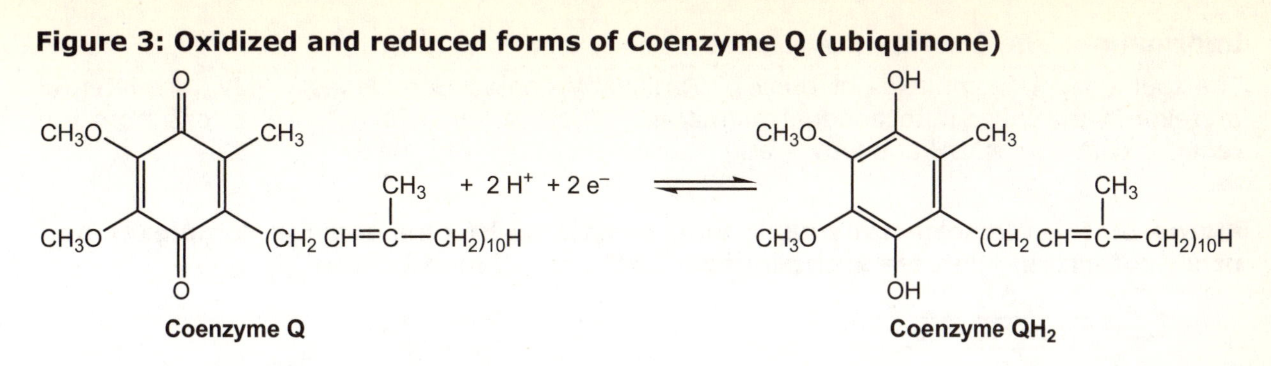

Figure 4: Oxidized and reduced forms of the heme group of a *b* cytochrome. Cytochromes are heme-containing proteins.

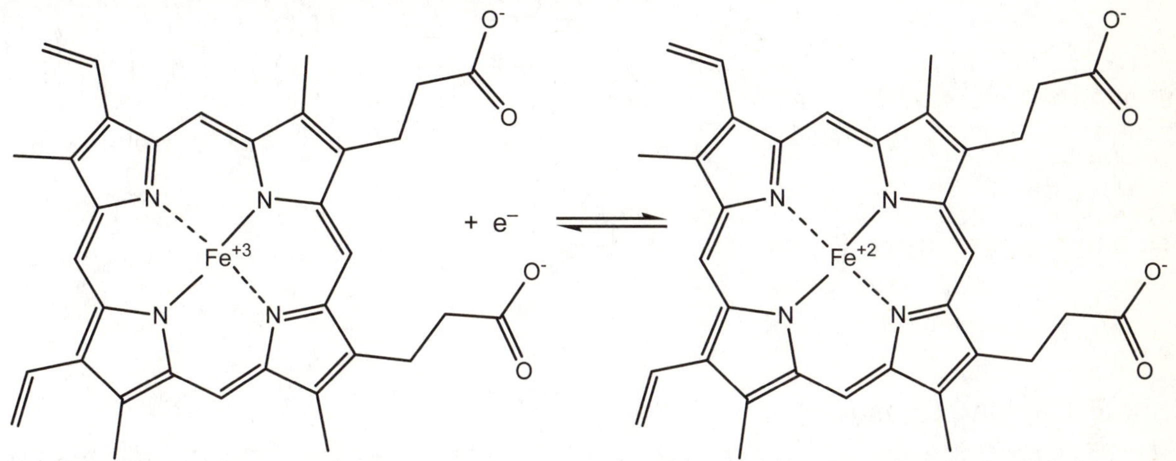

Critical Thinking Questions:

Refer to Figures 1-4 to help you answer CTQ's 1-6.

1. Which of the four types of electron carriers in the ETC are **one-electron** carriers?

2. Which of the four types of electron carriers in the ETC are **two-electron** carriers?

3. Which two electron carriers in the ETC are **proteins**? What metal ion cofactor is required for these carriers?

4. Which two electron carriers in the ETC are **coenzymes**?

5. Cytochromes are heme-containing proteins. What other common blood protein contains heme?

6. What is the important metal ion in heme?

Information: The Enzyme Complexes

The four enzyme complexes of the ETC (brilliantly named Complexes I – IV) are integral proteins in the inner mitochondrial membrane. They serve to take the electrons from the reduced cofactors NADH and $FADH_2$ and transfer them to O_2 to make H_2O.

Figure 5: The electron-transport chain, consisting of four enzyme complexes and other cofactors, operates in the inner mitochondrial membrane

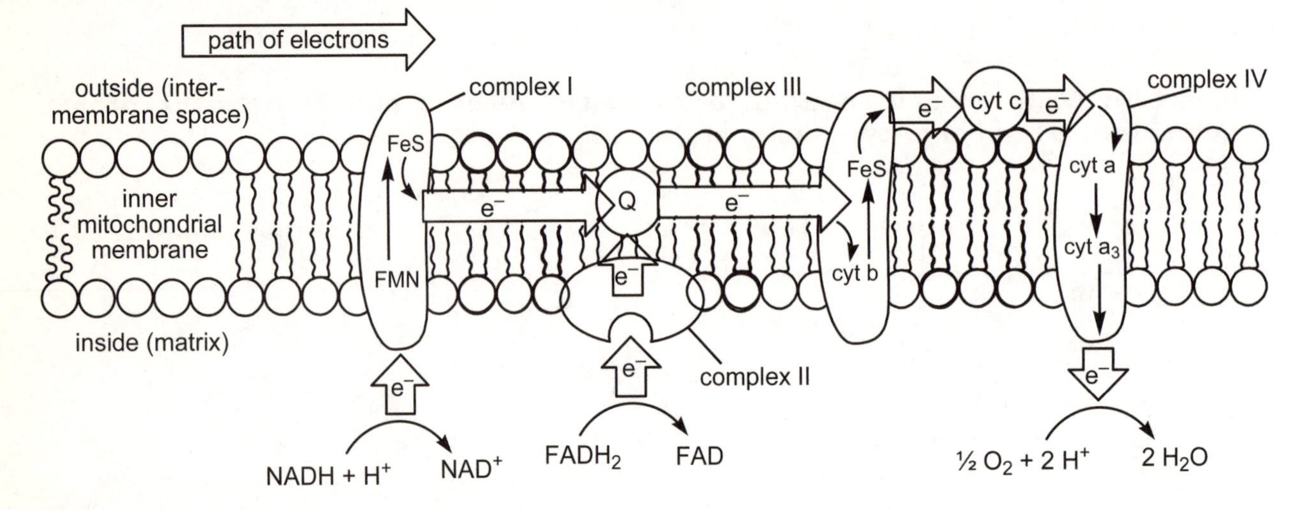

Critical Thinking Questions:

Refer to Figure 5 to help you answer CTQ's 7-13. The path of electrons through the various complexes and cofactors follows the hollow arrows (⇒).

7. Which complex accepts electrons from NADH?
8. Which complex accepts electrons from $FADH_2$?
9. Which complexes contain iron-sulfur clusters?
10. Which complexes contain cytochromes?
11. What molecule carries electrons from Complex I to Complex III?
12. What molecule carries electrons from Complex III to Complex IV?
13. Which complex transfers electrons to oxygen?
14. For electrons entering the ETC from NADH, which of the complexes I – IV do the electrons pass through? List them.

15. For electrons entering the ETC from NADH, which of the complexes I – IV do the electrons pass through? List them.

Information: Oxidative Phosphorylation

All the reactions of the ETC are energetically favorable, and "use up" the energy stored in the reduced cofactors. This energy must be saved somehow to make ATP. *The energy is saved by using the favorable electron-transport reactions to "pump" protons (H^+ ions) out of the matrix into the intermembrane space.* This "proton gradient" (also called "pH gradient") is a high-energy state, because the natural tendency is for ions to diffuse from areas of higher concentration to areas of lower concentration.

The only way for the protons to reenter the matrix is through an enzyme complex called ATP synthase. So, the protons are pumped out of the matrix, and as they return, providing energy like water turning a waterwheel, they are used to attach a phosphate to ADP to make ATP. This process is called **oxidative phosphorylation**, because the energy needed to phosphorylate ADP to make ATP comes from the oxidative steps of the ETC.

Figure 6: When electrons traverse Complexes I, III, and IV of the ETC, Protons are pumped out of the matrix. Protons reenter through the ATP synthase complex (V), providing energy to make ATP. Complex II (not shown) pumps no protons.

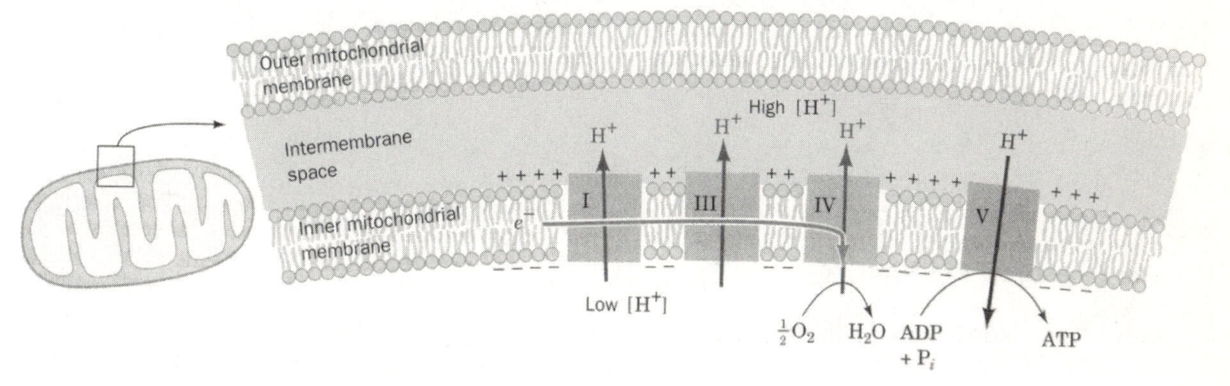

[From Biochemistry, 2E; Voet, D. and Voet, J.G.; Copyright © (1995), John Wiley & Sons, Inc. Reprinted with permission of John Wiley & Sons, Inc.]

Critical Thinking Questions:

16. Suppose that as two electrons proceed through the ETC as shown in Figure 6, Complexes I, III, and IV each pump **two** protons out of the matrix. Considering your answers to CTQs 14 and 15, how many total protons are pumped:

    a. For the two electrons originating in NADH?

    b. For the two electrons originating in $FADH_2$?

17. Considering your answer to CTQ 16, if the oxidation of one NADH to NAD^+ *via* the ETC leads to production of 3 ATP molecules *via* ATP synthase (Complex V), how many ATP could be produced from oxidation of one $FADH_2$? Explain your answer.

Exercises:

1. Complete the table below with the numbers of products that would be obtained in the various steps from the aerobic oxidation of glucose. Then total the ATP that one glucose is "worth."

	Glycolysis	"Bridge to CAC"	CAC	Total	# of ATP "worth"
ATP (or GTP)	2	0	2	4	4
NADH					
$FADH_2$					

2. Biochemists measure pH to determine the values given in CTQ 16 for the number of protons pumped out by each ETC enzyme complex. Explain how variations in these measurements would lead to a different answer to Exercise 1.

3. In Figure 2, the iron atoms are coordinated to four amino-acid residues in a protein. All four amino-acid residues are the same. Identify the amino acid.

4. Read the assigned pages in your text, and work the assigned problems.

Fatty Acid Oxidation

(How is energy produced from fats?)

Information:

Fats (triacylglycerols, Figure 1) are degraded during digestion to fatty acids and monoacylglycerols, but are reformed into triacylglycerols ("triglycerides") and packaged with protein into chylomicrons for transport in the bloodstream. After fats enter a cell, they again are broken down into fatty acids. Fatty acids are degraded in mitochondria to acetyl-CoA, which yields energy *via* the citric acid cycle and electron transport, in the same manner as we have seen previously.

Figure 1: Stearin, a triacylglycerol

Figure 2: Oxidation of fatty acids begins at the β carbon.

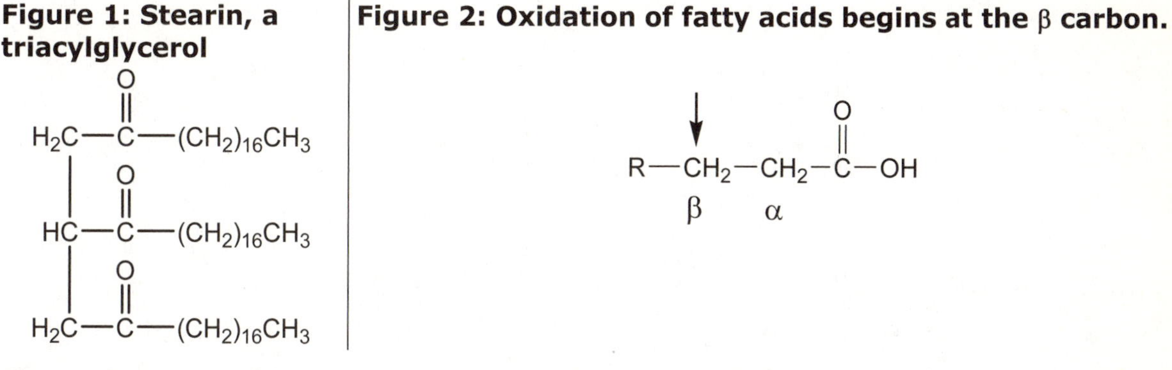

Figure 3: Fatty acid activation takes place in the cytosol.

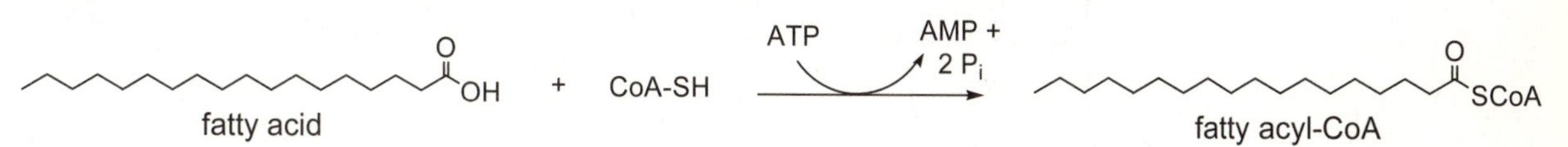

Fatty acid oxidation is usually called β-oxidation (beta-oxidation), since the β carbon of the fatty acid is oxidized in the process (see Figure 2).

The free fatty acid is first activated for oxidation by attaching it to coenzyme A in a process that requires ATP. Then, the activated fatty-acyl-CoA is transported into the mitochondrial matrix.

Critical Thinking Questions:

Refer to Figure 3 (above) and Figure 4 (following page) to help you answer CTQs 1-6.

1. How many "high-energy phosphate bonds" are hydrolyzed during fatty acid activation (Figure 3)? How can you tell?

2. Which of the four steps in the β-oxidation pathway are oxidations? How can you tell?

Figure 4: β-oxidation removes 2-carbon units from fatty-acyl-CoA to produce acetyl-CoA.

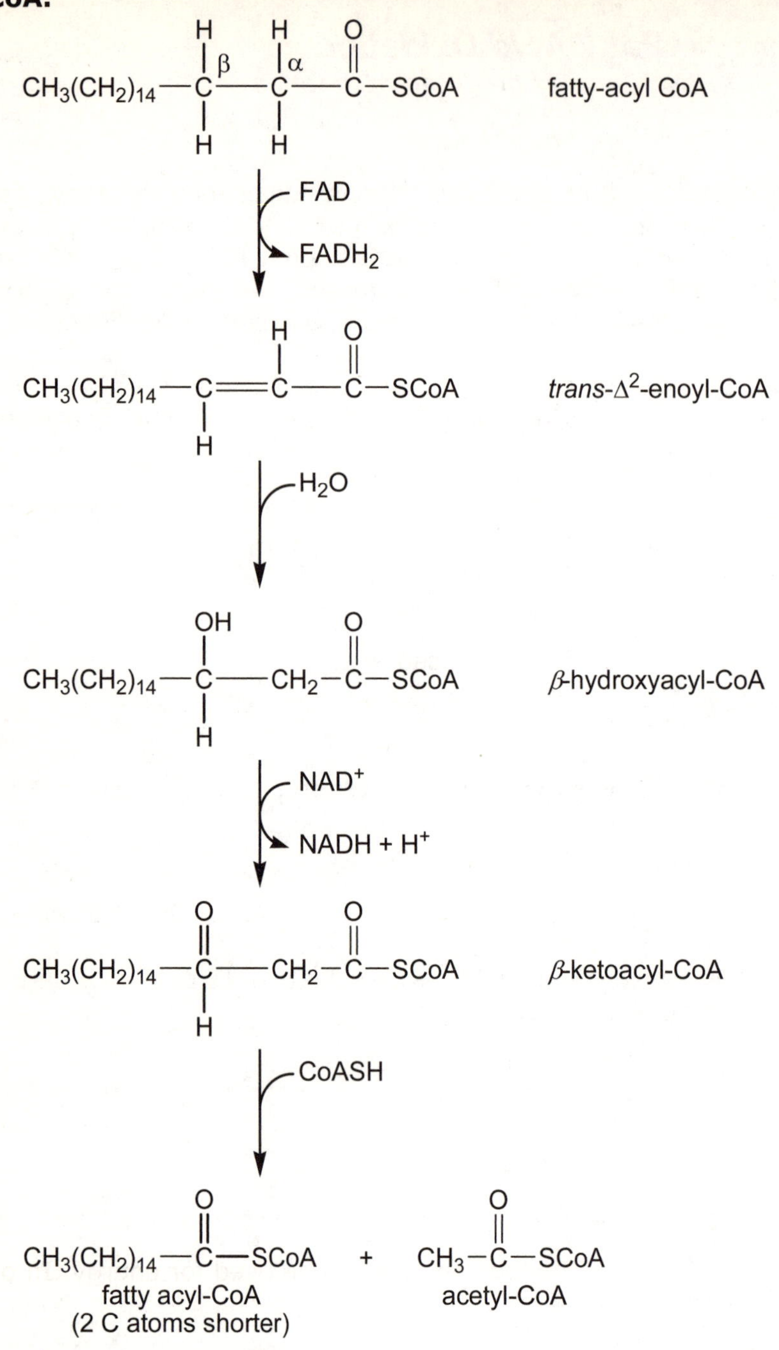

Critical Thinking Questions:

3. Which of the six classes of enzymes would be used in:
   a. Step 1 ____________________
   b. Step 2 ____________________
   c. Step 3 ____________________
   d. Step 4 ____________________

4. Suppose that step 1 of β-oxidation begins with an 18-carbon fatty-acyl-CoA, as in Figure 4. Step 4, then, produces acetyl-CoA and a 16-carbon fatty-acyl-CoA.

   a. What would happen next in order to oxidize the 16-carbon fatty-acyl-CoA?

   b. Explain why β-oxidation is said to be a spiral rather than a cyclic pathway.

5. Consider the β-oxidation of a C_{14} fatty-acyl-CoA.

   a. How many spirals of β-oxidation would be needed to completely degrade the C_{14} fatty-acyl-CoA to acetyl-CoA?

   b. Explain why the answer to part (a) is not 7.

   c. How many $FADH_2$ molecules would be produced? ____

   d. How many NADH molecules would be produced? ____

   e. How many acetyl-CoA molecules would be produced? ____

   f. Remembering that each acetyl-CoA can be oxidized in the citric acid cycle to produce 3 NADH, 1 $FADH_2$, and 1 GTP, total the number of ATP produced from the C_{14} fatty acid after complete aerobic oxidation, including electron transport. Don't forget to subtract the number that you gave in the answer to CTQ 1. Show your work.

Information: Ketone bodies

Acetyl-CoA must enter the citric acid cycle (CAC) in order to be used for energy. In order for this to happen, oxaloacetate (a carbohydrate) must be present to react with acetyl-CoA in step 1 of the CAC. During fasting, or if the diet is high in fat and low in carbohydrates, there is not sufficient oxaloacetate to react with, and so the concentration of acetyl-CoA builds up. When this happens, other enzymes catalyze a condensation of acetyl-CoA with itself to form blood-soluble molecules called ketone bodies (see Figure 5). These reactions are reversible, so ketone bodies may revert to acetyl-CoA and be used for energy when oxaloacetate is available.

If the concentration of ketone bodies becomes too high, as can happen in diabetics or on a very low carbohydrate diet, a condition called **ketoacidosis** can result. Since the ketone bodies are not completely metabolized, acetone can diffuse out of the bloodstream into air in the lungs. Also, since the ketone bodies acetoacetate and β-hydroxybutyrate are acidic, the pH of the blood decreases. This affects the ability of the hemoglobin to carry oxygen, and breathing can become difficult. Untreated, this condition can lead to coma, or even death.

Figure 5: An excess of acetyl-CoA leads to the formation of the "ketone bodies" acetoacetate, β-hydroxybutyrate, and acetone

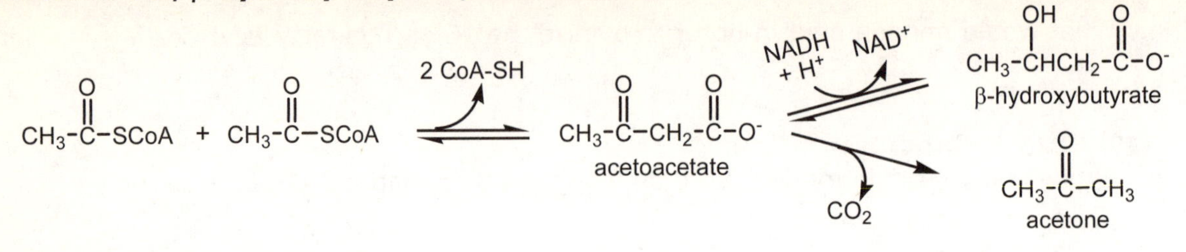

Critical Thinking Questions:

6. Ketone bodies can be removed by the kidneys and excreted in the urine. Would this increase or decrease the effectiveness of a low carbohydrate diet? Explain your answer.

7. A patient on a "low-carb" diet is being monitored by a physician. Why might the physician sniff the breath of the patient to check his health?

Exercises:

1. Consider stearic acid, an 18-carbon saturated fatty acid.
   a. Calculate the total number of ATP that could be produced by the complete oxidation of stearic acid.

   b. Calculate the ratio of the number of ATP formed per carbon atom in stearic acid.

   c. Considering the complete oxidation of glucose to yield 38 ATP, calculate the ratio of the number of ATP formed per carbon atom in glucose.

2. Carbons that are in more reduced oxidation states can release more energy when they are oxidized. On average, are the carbons in a fatty acid more reduced or more oxidized than those in a carbohydrate? Explain.

3. Read the assigned pages in your text, and work the assigned problems.

Other Metabolic Pathways

(In what other ways is energy produced or used?)

Information: Catabolism of Proteins

We have seen how carbohydrates (specifically, glucose) and fats can be used for energy. Proteins can be hydrolyzed into their amino acid components, which can be transformed into other energy-producing molecules as shown in Figure 1.

Figure 1: Carbons from amino acids are degraded to CAC intermediates and other related metabolites

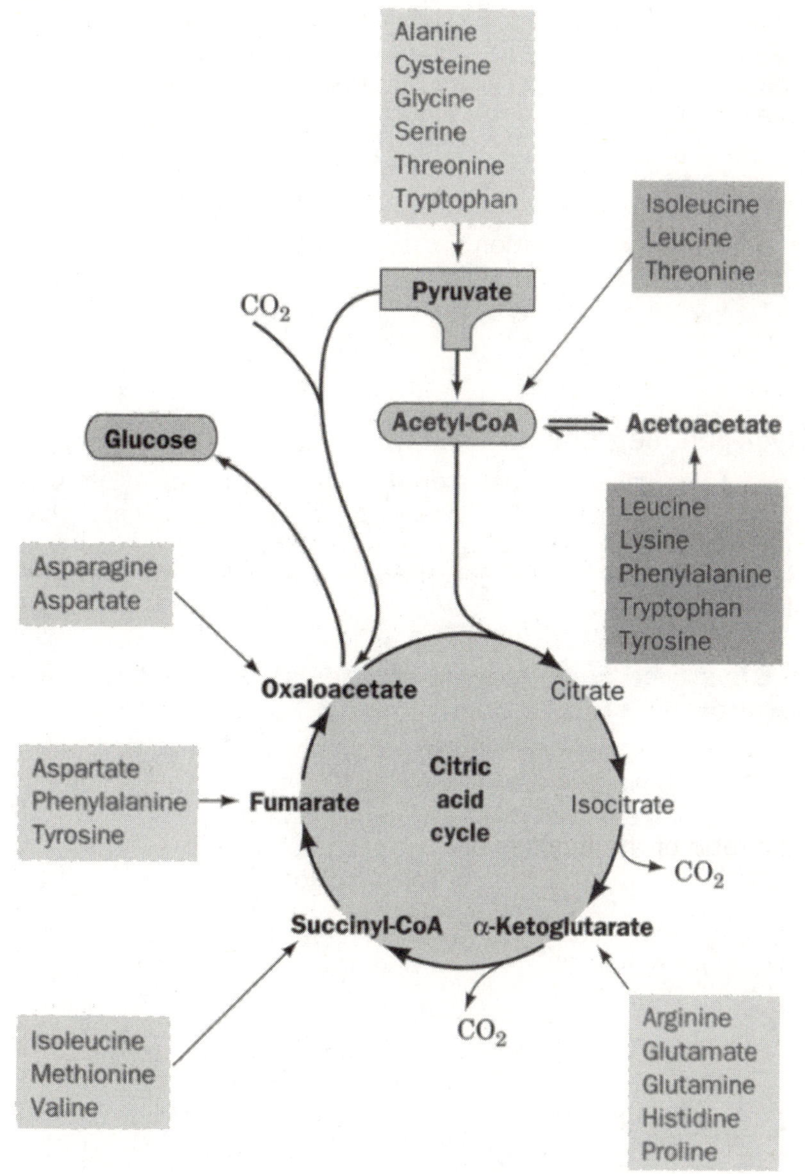

[From Biochemistry, 2E; Voet, D. and Voet, J.G.; Copyright © (1995), John Wiley & Sons, Inc. Reprinted with permission of John Wiley & Sons, Inc.]

Critical Thinking Questions:

1. Assuming that no ATP are used or formed in the transformation of the amino acid serine into pyruvate, how many ATP can be produced using serine for energy? Show your work. (Hint: Consider **all** the cofactors produced during its oxidation, -see ChemActivity 41, Figure 2-if necessary.)

2. Assuming that no ATP are used or formed in the transformation of the amino acid lysine into acetoacetyl-CoA, how many ATP can be produced using lysine for energy? Show your work. (See also ChemActivity 43, Figure 5.)

Information: Glycogen metabolism

Recall that glycogen (also called animal starch) is a polymer of α-D-glucose connected by α–1,4 linkages with α-1,6 branches. When energy is plentiful, excess glucose is converted into glycogen, much of which is stored in the liver. When energy is needed, the glycogen can undergo phosphorolysis to produce glucose-6-phosphate. These processes are summarized in Figure 2.

Figure 2: Glucose is stored as glycogen (synthesis), and released as glucose-1-phosphate (phosphorolysis)

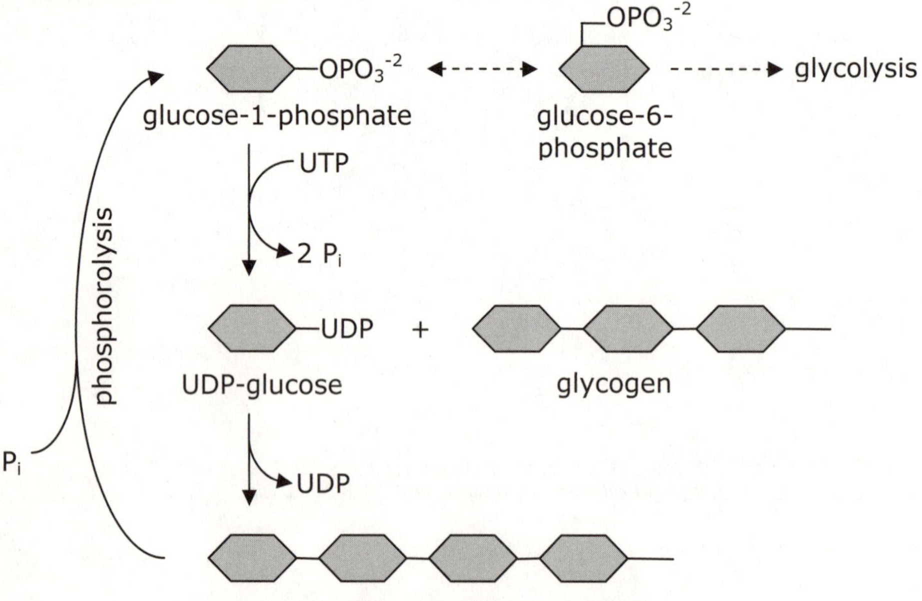

Critical Thinking Questions:

3. How many "high-energy phosphate bonds" are hydrolyzed in order to attach one glucose to glycogen? Explain how you can tell.

4. Explain why glycolysis starting from glycogen yields one more ATP than when starting with glucose itself. You may need to refer to Figure 1 of ChemActivity 40.

Information: Gluconeogenesis

The brain normally requires glucose as an energy source (although it can survive on ketone bodies when fasting, if required). When glycogen supplies run low, liver enzymes can perform a process called gluconeogenesis, which literally means "new birth of glucose."

Figure 3: Gluconeogenesis produces glucose when glycogen stores are depleted. Pyruvate and oxaloacetate (from the CAC) are starting materials, but acetyl-CoA is not.

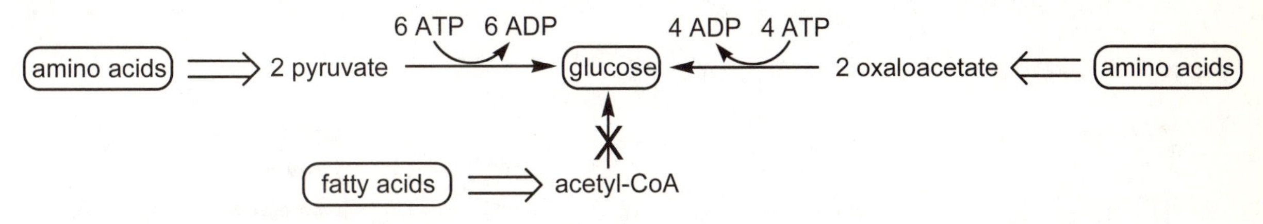

Critical Thinking Questions:

5. Recall that some amino acids break down to make citric acid cycle intermediates. Persons on very low calorie diets often lose muscle mass along with body fat. Explain why this is probably unavoidable.

6. Humans can survive poorly on a diet of protein with very little fat or carbohydrates, but not at all on a diet of fat with little protein or carbohydrates. Explain.

Information: Anabolism—Pentose phosphate pathway

Catabolic pathways produce energy, and are considered oxidative, but anabolic (biosynthetic) pathways require energy and are considered reductive. While oxidative pathways require oxidized cofactors such as NAD^+ and FAD, reductive pathways require cofactors in their reduced form.

The electron transport chain keeps most of the cofactors NADH and $FADH_2$ in their oxidized forms, so they are not available in their reduced forms for biosynthesis. Therefore, reductive pathways make use of different cofactors, such as NADPH. (By attaching a phosphate group to NADH, it becomes NADPH.)

The **pentose phosphate pathway** (also called the phosphogluconate pathway or hexose monophosphate shunt) produces NADPH for what is called "reducing power"—the ability to perform biosynthesis. This pathway can also be used nonoxidatively to produce ribose-5-phosphate, which is needed for the backbone structure of the nucleic acids RNA and DNA.

Figure 4: The pentose phosphate pathway produces ribose-5-phosphate and NADPH

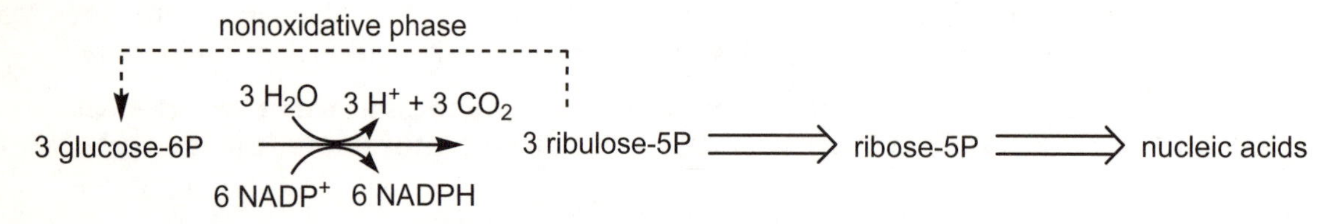

Figure 5: The three major uses of glucose

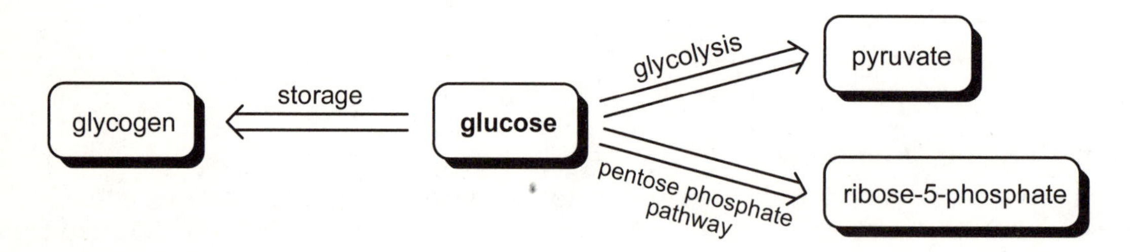

Critical Thinking Questions:

7. What are the two main products of the pentose phosphate pathway? For what purpose are the products used?

8. Of the three fates of glucose in Figure 4 (storage, PPP, or glycolysis), two occur in most cells at all times. The other one alternates on and off. Explain.

9. Which of the three pathways would increase in rate mainly during cell replication or division? Explain.

Information: Fatty acid synthesis

Fatty acid synthesis is a spiral pathway similar to the reverse of β-oxidation. Main differences are the location (synthesis takes place in the cytosol), and the coenzymes and cofactors used. Also, the fatty-acyl groups are bonded to an acyl carrier protein instead of to coenzyme A.

Figure 6: Fatty acid synthesis adds 2-carbon units from malonyl-ACP to the growing fatty-acyl-ACP. (ACP = Acyl Carrier Protein).

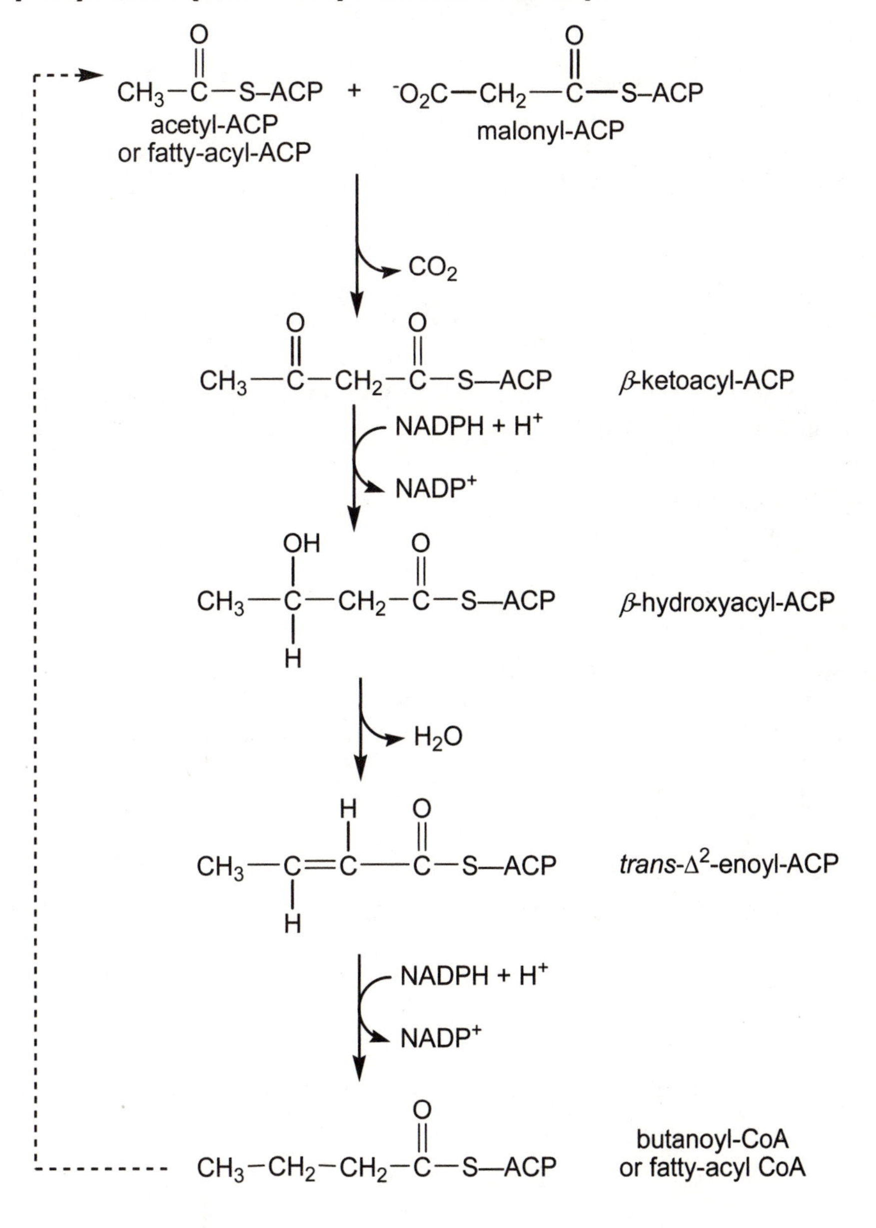

Critical Thinking Questions:

10. Which of the six classes of enzymes would catalyze each of the reactions shown in Figure 6 for fatty acid synthesis?

    a. Step 1 ____________________

    b. Step 2 ____________________

    c. Step 3 ____________________

    d. Step 4 ____________________

11. A high blood glucose level stimulates the release of the hormone insulin, which activates fatty acid biosynthesis. Why would this be an appropriate response of the organism?

Exercises:

1. Propose some reasons that fatty acid synthesis requires NADPH rather than NADH as a cofactor.

2. Glycolysis converts glucose to pyruvate, and gluconeogenesis converts pyruvate to glucose. The conversion of glucose → pyruvate → glucose by this path is called a futile cycle.

   a. Calculate the net number of ATP molecules that would be produced or required for this futile cycle.

   b. High blood glucose levels bring about the secretion of the hormone insulin. Insulin activates glycolysis and inactivates gluconeogenesis. Explain why it is beneficial for the organism to have both effects occur simultaneously.

   c. Would you expect insulin to activate or inactivate glycogen synthesis? Explain.

   d. Would you expect insulin to activate or inactivate glycogen phosphorolysis? Explain.

3. Read the assigned pages in your text, and work the assigned problems.

Stoichiometry (Mole Relationships): Practice Worksheet 1

1. Given the equation for the combustion of butane: $2\ C_4H_{10} + 13\ O_2 \rightarrow 8\ CO_2 + 10\ H_2O$
   Show what the following molar ratios should be.
   a. C_4H_{10} / O_2

   b. O_2 / CO_2

   c. O_2 / H_2O

   d. C_4H_{10} / CO_2

   e. C_4H_{10} / H_2O

2. Given the reaction: $4\ NH_3\ (g) + 5\ O_2\ (g) \rightarrow 4\ NO\ (g) + 6\ H_2O\ (l)$

   When 1.20 mol of ammonia reacts, the total number of moles of products formed is:

   a. 1.20 b. 1.50 c. 1.80 d. 3.00 e. 12.0

3. Silver sulfide (Ag_2S) is the common tarnish on silver objects.
   a. What weight of silver sulfide can be made from 1.23 mg of hydrogen sulfide (H_2S) obtained from a rotten egg?
   b. It 5.7 mg of silver sulfide are obtained, what is the percent yield?

   The reaction of formation of silver sulfide is:

   $Ag(s) + H_2S(g) + O_2(g) \rightarrow Ag_2S(s) + H_2O(l)$ (Equation must first be balanced.)

4. Given the following balanced equation: $2\ KClO_3 \rightarrow 2\ KCl + 3\ O_2$

   How many moles of O_2 can be produced by letting 12.00 moles of $KClO_3$ react?

5. A somewhat antiquated method for preparing chlorine gas involves heating hydrochloric acid with pyrolusite (manganese dioxide), a common manganese ore. (Reaction given below). How many kg of HCl react with 5.69 kg of manganese dioxide?

   $HCl(aq) + MnO_2(s) \rightarrow H_2O(l) + MnCl_2(aq) + Cl_2(g)$ (Equation must first be balanced.)

6. Given the following equation: $2\ K + Cl_2 \rightarrow 2\ KCl$
   a. How many grams of KCl are produced from 2.50 g of K and 3.00 g Cl_2?

   b. How many grams of KCl are produced from 1.00 g of Cl_2 and excess K?

7. Given the following equation: $Na_2O + H_2O \rightarrow 2\ NaOH$
   a. How many grams of NaOH are produced from 1.20×10^2 grams of Na_2O?

   b. How many grams of Na_2O are required to produce 1.60×10^2 grams of NaOH?

8. Given the following equation: 8 Fe + S_8 → 8 FeS
   a. What mass of iron is needed to react with 16.0 grams of sulfur?

   b. How many grams of FeS are produced?

9. Given the following equation: 2 $NaClO_3$ → 2 NaCl + 3 O_2
   a. 12.00 moles of $NaClO_3$ will produce how many grams of O_2?

   b. How many grams of NaCl are produced when 80.0 grams of O_2 are produced?

10. Given the following equation: Cu + 2 $AgNO_3$ → $Cu(NO_3)_2$ + 2 Ag
    a. How many moles of Cu are needed to react with 3.50 moles of $AgNO_3$?

    b. If 89.5 grams of Ag were produced, how many grams of Cu reacted?

11. Molten iron and carbon monoxide are produced in a blast furnace by the reaction of iron(III) oxide and coke (pure carbon). If 25.0 kilograms of pure Fe_2O_3 and 100.0 kilograms of pure carbon are used, how many kilograms of iron can be produced? The reaction is:

$$Fe_2O_3 + 3\ C \rightarrow 2\ Fe + 3\ CO$$

12. The "average human" requires 120.0 grams of glucose ($C_6H_{12}O_6$) per day. How many grams of CO_2 (in the photosynthesis reaction) are required for this amount of glucose? The photosynthetic reaction is:

$$6\ CO_2 + 6\ H_2O \rightarrow C_6H_{12}O_6 + 6\ O_2$$

Answers to Stoichiometry Practice:

1. a. 2/13 b. 13/8 c. 13/10 d. 2/8 e. 2/10
2. The correct answer is d.

   $NH_3 \rightarrow (NO + H_2O) = 4 \rightarrow 10$

   $4 / 10 = 1.20 / x$

   $x = 3.00$ mol
3. a. 8.95 mg Ag_2S b. 64%
4. 18.00 mol O_2
5. 9.54 kg HCl
6. a. 4.77 g KCl b. 2.10 g KCl
7. a. 155 g NaOH b. 124 g Na_2O
8. a. 27.9 g Fe b. 43.9 g FeS
9. a. 576.0 g O_2 b. 97.4 g NaCl
10. a. 1.75 mol Cu b. 26.4 g Cu
11. 17.5 kg Fe
12. 175.9 g CO_2

Gases: Practice Worksheet

Information: $R = 0.0821 \frac{L \cdot atm}{K \cdot mol}$ 1 atm = 760 torr = 760 mm Hg

Suggested demonstration: How to solve gas law problems

A. Simple substitution **B. Changing conditions**

1. A temperature of 0°C and a pressure of 1 atm are defined as the *Standard Temperature and Pressure*, or *STP*, for calculations involving gases. Use the ideal gas law to calculate the volume occupied by one mole of air at STP.

2. A balloon filled with air at room temperature (22°C) contains 5.0 L of air. What volume would the balloon occupy if it were cooled to liquid nitrogen temperature (–196°C)?

3. What is the density of methane gas (natural gas), CH_4, at 25°C and 0.947 atm? Express your answer in **grams per liter**. (Hint: d=m/V; assume any V you like, and calculate its mass).

4. When a 2.0-L bottle of concentrated HCl was spilled, 1.2 kg of $CaCO_3$ was required to neutralize the spill. Considering the reaction equation shown, what volume of CO_2 was released by the neutralization at 735 mm Hg and 20°C? (Hint: besides V, what variable in the ideal gas law is not given?)

   $CaCO_3(s)$ + 2 $HCl(aq)$ ssd $CaCl_2(aq)$ + $H_2O(l)$ + $CO_2(g)$

5. The partial pressure of air in the alveoli, the air sacs in the lungs, was measured as follows: nitrogen, 570.0 mm Hg; carbon dioxide, 40.0 mm Hg; and water vapor, 47.0 mm Hg. If the barometric pressure was 0.973 atm, what was the pO_2 in the alveoli? Express your answer in millimeters of mercury (torr).

Answers: 1. 22.41 L 2. 1.3 L 3. 0.621 g/L 4. 3.0×10^2 L 5. 82.5 mm Hg or 82.5 torr

Additional Exercises:

1. The density of dry air at 25°C is about 1.2 kg/m^3. If you open a natural gas (methane) spigot in the chemistry laboratory, will the methane sink to the floor or rise toward the ceiling? (Convert units and compare with #3 above).
2. Read the assigned pages in your textbook, and work the assigned problems.

Stoichiometry (Mole Relationships): Practice Worksheet 2

Information:

Recall that an acid and a base react to form a salt and water. This is called a neutralization reaction. If we put a pH indicator in a solution of acid and then add base until the solution is neutralized, the indicator changes color. This *indicates* that we have added equal moles of acid and base, and the *endpoint* has been reached. This process is called a *titration*.

Consider the reaction of sulfuric acid and sodium hydroxide:

$$H_2SO_4\ (aq) + 2\ NaOH\ (aq) \rightarrow 2\ H_2O\ (l) + Na_2SO_4\ (aq)$$

We would have to add 2 moles of NaOH to neutralize 1 mole of H_2SO_4, giving mole conversion factors of:

$$\frac{2\text{ mol NaOH}}{1\text{ mol } H_2SO_4} \quad \text{or} \quad \frac{1\text{ mol } H_2SO_4}{2\text{ mol NaOH}}$$

To determine the moles of base (NaOH), we need to know both the **amount** (mL or L) of the base and the **concentration** (M or mol/L) of the base that was added. We can then use the mole ratio to determine how much acid is present.

The **unit plan** would be: L NaOH $\rightarrow$ mol NaOH $\rightarrow$ mol H_2SO_4

Once you calculate the mol H_2SO_4, then you may simply divide by the volume that was used to get mol/L (*i. e.,* M).

Exercises:

1. When 25.00 mL of a sulfuric acid solution of unknown concentration was titrated with sodium hydroxide solution, the acid required 42.36 mL of 0.500 M NaOH to reach an endpoint.

    a. Write a unit plan for the titration (stoichiometry) problem to calculate the moles of acid in the sample. Then write conversion factors for each step.

    b. How many moles of sulfuric acid were present? Show work with all units.

    c. What was the concentration (in M) of the sulfuric acid?

d. What was the concentration in % (m:v)? Make a unit plan first.

2. In a titration, a 20.00-mL portion of Lime-a-Way cleaner containing the monoprotic acid sulfamic acid (HSO_3NH_2) required 30.26 mL of 0.102 M NaOH to reach an endpoint.

   a. Write a balanced equation for the reaction of sulfamic acid and sodium hydroxide.

   b. Write the two mole factors for the titration.

   c. Write a unit plan for the titration (stoichiometry) problem.

   d. Calculate the concentration of sulfamic acid in Lime-a-Way in molarity.

   e. What is the concentration of sulfamic acid in % (m:v)? Write a unit plan first.

Answers: 1. b. 1.06×10^{-2} mol c. 0.424 M d. 4.15 % 2. d. 0.154 M e. 1.50 %

Functional Groups

(How are organic molecules classified?)

Information:

Organic molecules can be thought of in terms of **functional groups—**certain arrangements of bonded atoms that have predictable properties and reactivities.

The tables below list some functional groups that are important in organic chemistry and biochemistry. Table 1 contains those which we will name by the systematic rules agreed upon by the International Union of Pure and Applied Chemistry (IUPAC). These rules assume that the most important functional group is assigned a **base name**, and other functional groups are named as **substituents**. Most functional groups have separate rules depending on whether it is the base name (named last) or the substituent (named first); these differences are indicated in the columns "base suffix" and "name when substituent" in the table.

Table 2 contains some functional groups which are easier to name by common (or trivial) naming rules.

Sometimes more than one name is given for an example because IUPAC has adopted some common names as acceptable alternates. For example, the molecule *formaldehyde* is systematically named *methanal*. The second (systematic) name is more descriptive, but almost nobody uses it. You should be familiar with both names.

Table 1: Organic Functional Groups (Systematic IUPAC Naming Rules)

functional group	generic structure	base suffix	name when substituent	examples	
alkane	R	-ane	-yl	$CH_3CH(CH_3)CH_3$ 2-methylpropane	isopropylcyclohexane
alkene	C=C	-ene	-enyl	*trans*-2-pentene	4-bromocyclopentane
alkyne	—C≡C—	-yne	-ynyl	$CH_3C{\equiv}C{-}CH(CH_3)CH_3$ 4-methyl-2-pentyne	
aromatic	(benzene ring)	benzene	phenyl-	Cl–C₆H₄–Cl 1,4-dichlorobenzene	2-phenylheptane
haloalkane	R-X	—	fluoro-, chloro- *etc.*	CF_2Cl_2 dichlorodifluoromethane	Br, Br *trans*-1,2-dibromocyclopentane

Table 1 (continued): Organic Functional Groups (Systematic IUPAC Naming Rules)

functional group	generic structure	base suffix	name when substituent	examples		
alcohol	R-OH	-ol	hydroxy-	$CH_3CH(OH)CH_3$ 2-propanol	$CH_3CH(OH)C(=O)-OH$ 2-hydroxypropanoic acid	
phenol	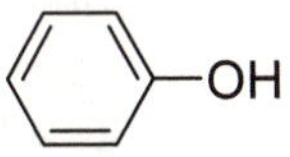 C_6H_5-OH	phenol	—	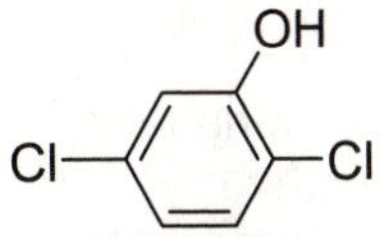 2,5-dichlorophenol		
thiol	R-SH	-thiol	mercapto-	HS-$CH_2CH_2CH_3$ 1-propanethiol	$HSCH_2CH_2OH$ 2-mercaptoethanol	
aldehyde	R-C(=O)-H	-al	—	H-C(=O)-H methanal (formaldehyde)	C_6H_5-C(=O)-H benzaldehyde	$CH_3CH(Cl)CH(=O)$ 2-chloro-propanal
ketone	R-C(=O)-R	-one	—	CH_3-C(=O)-CH_3 2-propanone (acetone)	CH_3-C(=O)-$CH(CH_3)CH_3$ 3-methyl-2-butanone (methyl isopropyl ketone)	
carboxylic acid	R-C(=O)-OH	-oic acid	—	CH_3-C(=O)-OH ethanoic acid (acetic acid)	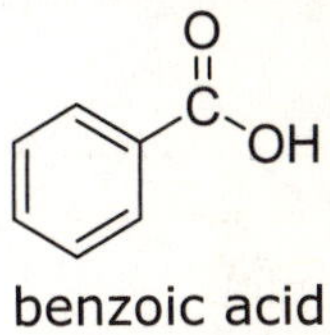 C_6H_5-C(=O)-OH benzoic acid	
acid salt	R-C(=O)-$O^{\ominus}$	-oate	—	CH_3-C(=O)-$O^{\ominus}$ $Na^{\oplus}$ sodium ethanoate (sodium acetate)	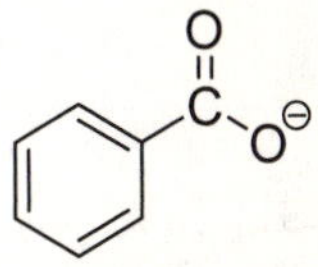 C_6H_5-C(=O)-$O^{\ominus}$ benzoate	
amide	R-C(=O)-NR_2	-amide	—	CH_3-C(=O)-NH_2 ethanamide (acetamide)	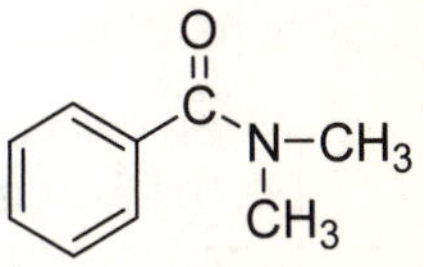 C_6H_5-C(=O)-N(CH_3)-CH_3 N,N-dimethylbenzamide	

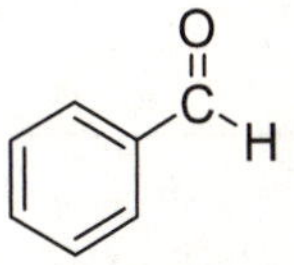

Table 2: Organic Functional Groups (Common Naming Rules)

functional group	generic structure	naming rules	examples	
ester	$R-C(=O)-OR'$	name R′ as alkyl group; then like acid salt	$CH_3CH_2C(=O)-OCH_3$ methyl propanoate	$C_6H_5-C(=O)-O-CH(CH_3)CH_3$ isopropyl benzoate
acid anhydride	$R-C(=O)-O-C(=O)-R$	name each $R-C(=O)-$ as acid alphabetically; then "anhydride"	$CH_3CH_2C(=O)-O-C(=O)-CH_2CH_2CH_3$ butanoic propanoic anhydride	$CH_3-C(=O)-O-C(=O)-CH_3$ ethanoic anhydride (acetic anhydride)
ether	R-O-R	name each R as alkyl group alphabetically; then "ether"	$CH_3CH_2OCH_2CH_3$ diethyl ether	$C_6H_5-OCH_3$ methyl phenyl ether $C_6H_5-CH_2OCH_3$ benzyl methyl ether
sulfide	R-S-R	name each R as alkyl group alphabetically; then "sulfide"	$CH_3CH_2SCH_2CH_3$ diethyl sulfide	
disulfide	R-S-S-R	name each R as alkyl group alphabetically; then "sulfide"	$CH_3CH_2SSCH_2CH_3$ diethyl disulfide	
amine	$R-\ddot{N}(R)-R$	name each R as alkyl group alphabetically; then "amine"	CH_3NH_2 methyl amine	$(CH_3)_2NCH_2CH_3$ ethyl dimethyl amine
amino group	$-NH_2$	amino- prefix	$CH_3CH(NH_2)C(=O)-OH$ 2-aminopropanoic acid	$HOCH_2CH(NH_2)-CH_3$ 2-amino-1-propanol

Notes:

- Several functional groups contain a carbonyl group ($-C(=O)-$); the carbonyl is not considered a functional group by itself.
- When two or more functional groups are present, choose the more important one to be the **base name**. In general, the more carbon-oxygen bonds contained by a functional group, the higher its importance.

Exercises:

1. Identify the functional group that corresponds to the following descriptions.
   a. A carbonyl group connected to at least one hydrogen.
   b. An oxygen bonded to two saturated carbons.
   c. A benzene ring bonded to a hydroxy group.
   d. A carbon bonded to two other carbons and double-bonded to an oxygen.
2. List all the functional groups (other than alkane) in each of the following structures.

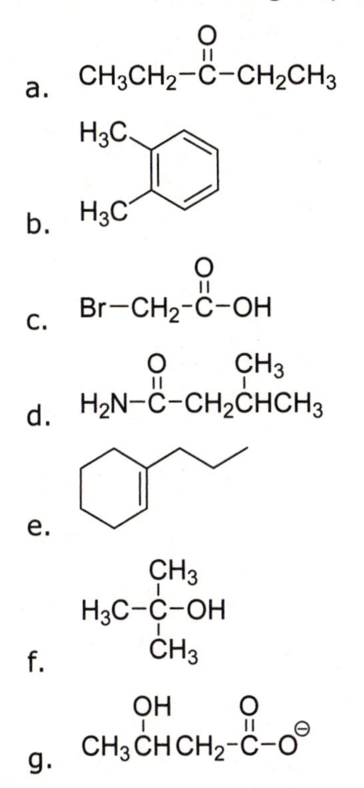

3. Give a correct (common or systematic) name for each molecule in Exercise 2.
   a.
   b.
   c.
   d.
   e.
   f.
   g.
4. Draw the structure that corresponds to each of the following names.
   a. diisopropyl disulfide

b. butyl methyl ether

c. 2-methyl-2-aminopropanal

d. benzoic anhydride

e. 2,4,6-trimethylphenol

f. pentyl amine

g. 2-mercaptoacetic acid

5. List all the functional groups (other than alkane) in each molecule named in Exercise 4.

a.

b.

c.

d.

e.

f.

g.

6. Read the assigned sections in your text and work the assigned problems.

Periodic Table of the Elements

1	2	3	4	5	6	7	8	9	10	11	12	13	14	15	16	17	18
1 **H** 1.008																	2 **He** 4.003
3 **Li** 6.941	4 **Be** 9.012											5 **B** 10.81	6 **C** 12.01	7 **N** 14.01	8 **O** 16.00	9 **F** 19.00	10 **Ne** 20.18
11 **Na** 22.99	12 **Mg** 24.31											13 **Al** 26.98	14 **Si** 28.09	15 **P** 30.97	16 **S** 32.07	17 **Cl** 35.45	18 **Ar** 39.95
19 **K** 39.10	20 **Ca** 40.08	21 **Sc** 44.96	22 **Ti** 47.88	23 **V** 50.94	24 **Cr** 52.00	25 **Mn** 54.94	26 **Fe** 55.85	27 **Co** 58.93	28 **Ni** 58.69	29 **Cu** 63.55	30 **Zn** 65.39	31 **Ga** 69.72	32 **Ge** 72.61	33 **As** 74.92	34 **Se** 78.96	35 **Br** 79.90	36 **Kr** 83.80
37 **Rb** 85.47	38 **Sr** 87.62	39 **Y** 88.91	40 **Zr** 91.22	41 **Nb** 92.91	42 **Mo** 95.94	43 **Tc** (98)	44 **Ru** 101.1	45 **Rh** 102.9	46 **Pd** 106.4	47 **Ag** 107.9	48 **Cd** 112.4	49 **In** 114.8	50 **Sn** 118.7	51 **Sb** 121.8	52 **Te** 127.6	53 **I** 126.9	54 **Xe** 131.3
55 **Cs** 132.9	56 **Ba** 137.3	57 **La** 138.9	72 **Hf** 178.5	73 **Ta** 180.9	74 **W** 183.9	75 **Re** 186.2	76 **Os** 190.2	77 **Ir** 192.2	78 **Pt** 195.1	79 **Au** 197.0	80 **Hg** 200.6	81 **Tl** 204.4	82 **Pb** 207.2	83 **Bi** 209.0	84 **Po** (209)	85 **At** (210)	86 **Rn** (222)
87 **Fr** (223)	88 **Ra** 226.0	89 **Ac** 227.0	104 **Rf** (261)	105 **Db** (262)	106 **Sg** (263)	107 **Bh** (262)	108 **Hs** (265)	109 **Mt** (268)	110 **Ds** (269)	111 (272)	112 (277)		114 (283)		116 (289)		
		58 **Ce** 140.1	59 **Pr** 140.9	60 **Nd** 144.2	61 **Pm** (145)	62 **Sm** 150.4	63 **Eu** 152.0	64 **Gd** 157.3	65 **Tb** 158.9	66 **Dy** 162.5	67 **Ho** 164.9	68 **Er** 167.3	69 **Tm** 168.9	70 **Yb** 173.0	71 **Lu** 175.0		
		90 **Th** 232.0	91 **Pa** 231.0	92 **U** 238.0	93 **Np** 237.0	94 **Pu** (244)	95 **Am** (243)	96 **Cm** (247)	97 **Bk** (247)	98 **Cf** (251)	99 **Es** (252)	100 **Fm** (257)	101 **Md** (258)	102 **No** (259)	103 **Lr** (260)